Henry Laing

Supplemental Descriptive Catalogue of Ancient Scottish Seals,

royal, baronial, ecclesiastical, and municipal, embracing the period from A.D. 1150

to the eighteenth century. Taken from original charters, and other deeds preserved

in public and private

Henry Laing

Supplemental Descriptive Catalogue of Ancient Scottish Seals,
royal, baronial, ecclesiastical, and municipal, embracing the period from A.D. 1150 to the eighteenth century. Taken from original charters, and other deeds preserved in public and private

ISBN/EAN: 9783337243944

Printed in Europe, USA, Canada, Australia, Japan

Cover: Foto ©berggeist007 / pixelio.de

More available books at **www.hansebooks.com**

SUPPLEMENTAL DESCRIPTIVE CATALOGUE

OF

ANCIENT SCOTTISH SEALS

Royal, Baronial, Ecclesiastical, and Municipal

EMBRACING THE PERIOD FROM A.D. 1150 TO THE EIGHTEENTH CENTURY

TAKEN FROM ORIGINAL CHARTERS AND OTHER DEEDS PRESERVED IN
PUBLIC AND PRIVATE ARCHIVES

BY HENRY LAING, EDINBURGH.

EDINBURGH
EDMONSTON AND DOUGLAS
1866.

TO THE

PRESIDENT AND MEMBERS

OF

THE SOCIETY OF ANTIQUARIES OF SCOTLAND

AND TO

THE OTHER SUBSCRIBERS

THIS VOLUME IS RESPECTFULLY DEDICATED

BY

HENRY LAING.

LIST OF SUBSCRIBERS.

Those marked with an Asterisk are dead.

The Duke of Argyll.
Society of Antiquaries of Scotland.
Society of Antiquaries, London.
Archæological Institute of Great Britain.
Royal Irish Academy.
Captain James H. Laurence-Archer, 60th Rifles.

The Duke of Buccleuch and Queensberry.
The Lord Belhaven and Stenton.
Sir Charles W. Blunt, of Heathfield Park, Baronet.
The Bishop of Brechin.
Charles Baker, Sackville Street, London.
Rev. W. K. R. Bedford, Sutton Coldfield, Warwickshire.
Rev. J. Blackburne, Leamington.
Messrs. Blackie and Son, Publishers, Glasgow.
James Blair, Rampart Row, Bombay (2 copies).
Edward Blore, F.S.A., D.C.L., Manchester Square, London.
Rev. Charles Boutell, Norwood, London.
Alexander J. Dennistoun Brown, of Balloch Castle.
George Burnett, Lyon-Depute, Edinburgh.

The Earl of Cawdor.
Walter Riddell-Carre, of Cavers-Carre.
John Inglis Chalmers, of Aldbar.
W. H. Clarke, York.
John Gilchrist Clark, of Speddoch.
Ebenezer Colquhoun, West George Street, Glasgow.
Thomas Constable, Printer to the Queen, Edinburgh (2 copies).
John Ross Coulthart, Croft House, Ashton-under-Lyne.
David Cowan, Writer to the Signet, Moray Place.
James T. Gibson Craig, Edinburgh (5 copies).
Raikes Currie, Minley Manor, Farnborough, Hants.

viii LIST OF SUBSCRIBERS.

The Earl of Dalhousie.
Sir George Scott Douglas, of Springwood Park, Baronet.
Sir John Douglas, K.C.B., Adjutant-General of H.M. Forces in Scotland.
The Dean and Chapter of Durham.
Rev. George Henry Dashwood, Vicar of Stow-Bardolph, Norfolk.
Neil Douglas C. F. Douglas, Scots Fusilier Guards.
George Stirling Home Drummond, Younger of Blair-Drummond.
Robert Dundas of Arniston.
William Pitt Dundas, Registrar-General of Scotland.
John Dundas, C.S., Edinburgh.

The Lord Elphinstone.
Walter Elliot, of Wolfelee, Hawick.
John M‘Dowell Elliot, Captain H.M. 4th King's Own Regiment.
William Euing, West George Street, Glasgow.
George Montgomery Ewan, Edinburgh.

The Lord Farnham.
Robert Fergusson, Morton, Carlisle.
Alexander Kinloch Forbes, Judge of the High-Court, Surat, Bombay.
Frederick Prescott Forteath, Captain H.M. 12th N.I., Bombay.
Augustus W. Franks, British Museum (2 copies).
Patrick Allan Fraser, Hospitalfield.
William Fraser, Assistant-Keeper of Sasines, Edinburgh.
W. N. Fraser, Solicitor Supreme Courts, Albany Street, Edinburgh.
Gilbert J. French, Bolton.

The Earl of Gosford.
Sir George Macpherson Grant, of Ballindalloch, Baronet.
Rev. John M. Gresley, Over Seile, Ashby-de-la-Zouch.

The Earl of Home.
John Buchanan Hamilton, of Leny and Bardowie.
Rev. J. Hannah, Warden of Trinity College, Glenalmond.
Robert Hay, of Linplum.
Isaac Anderson Henry, of Woodend.
Hugh Hopkins, Bookseller, Glasgow.
Rev. James Henry Hughes, M.A., Chaplain H.M. Bombay Ecclesiastical Establishment, Bombay (4 copies).
Rev. Eneas B. Hutchison, Incumbent of St. James's, Devonport.

Cosmo Innes, Professor of History, University of Edinburgh (2 copies).

Lord Jerviswoode.
Edward James Jackson, Coates Crescent, Edinburgh.
Andrew Jervise, Registration Examiner, Brechin.

The Earl of Kinnoull, Lord Lyon King of Arms.
Dr. J. Kendrick, Warrington, Lancashire.

The Lord Lindsay.
David Laing, Signet Library, Edinburgh.
John Bailley Langhorne, Wakefield.
Robert Laird, Clarenceux, College of Arms, London (2 copies).
Charles Lawson, Jun., George Square, Edinburgh.
George Logan, Teind Office, New Register House.

The Earl of Minto.
Sir John Maxwell, of Pollok, Baronet.
Sir William Stirling Maxwell, of Pollok, Baronet, M.P. (2 copies).
Sir G. Graham Montgomery, of Stanhope, Baronet, M.P.
Keith Stewart Mackenzie, of Seaforth.
John Whitefoord Mackenzie, Writer to the Signet, Edinburgh.
A. C. Mackenzie, of Findon.
Andrew Macgeorge, West Regent Street, Glasgow.
Graeme R. Mercer of Gorthy.
Anthony Morrison, Malabar Hill, Bombay (2 copies).

Mark Napier, Sheriff of Dumfriesshire, Edinburgh.
John Gough Nichols, Parliament Street, Westminster.

A. Haldane Oswald, of Auchincruive.

The Earl of Powis.
Charles S. Perceval, LL.D., Old Square, Lincoln's Inn, London.

H.M. General Register House.
Rev. James Raine, York.
Joseph Robertson, H.M. General Register House.
James C. Regor, Mincing Lane, London.

The Earl of Southesk.
Sir James Y. Simpson, Baronet, Edinburgh.

LIST OF SUBSCRIBERS.

PREFACE.

In the present day, when the science of Heraldry is receiving the attention which it may justly claim, and when, by the labours of such learned and accomplished writers as Nichols, Seton, Planché, Lower, Boutell, and others, the public mind seems disposed to receive any illustration of, or addition to, the knowledge of the " noble science," the author feels it to be comparatively an easier task to solicit a favourable reception for his present work than when he presumed to offer to the public the former " Descriptive Catalogue of Ancient Scottish Seals," of which this volume is a continuation.

Notwithstanding these favourable circumstances, however, he cannot but entertain considerable fears and misgivings in again venturing to appear before the public, conscious as he is of many shortcomings, and feeling deeply the disadvantages under which he labours. But he can say, with confidence, that he has made every exertion in his power, having spared neither time, labour, nor expense, to produce a work equal in all respects to the former volume, which, he may be allowed to remark, has been acknowledged by a competent authority to be " a valuable contribution to Scottish Heraldry ;" and he will be quite content if the present one shall prove equally useful. He has merely endeavoured to select and gather, from " hidden nooks and corners," materials for the use of more skilful hands, and to such he confidently leaves them.

Heraldry and Genealogy, when confined to their legitimate position, and divested of the absurd vagaries and fictions which have exposed them to the ridicule of wits and satirists, are undoubtedly invaluable auxiliaries to history : and in proportion as the student is acquainted with their principles and practice does he find his interest increased and his labours lightened. Nor is it to him alone that the advantage of such knowledge is confined. To all, indeed, who aspire to the distinction of having filled well their part in life, it will be found that the desire to "add virtue to their parentage" derives no little support and stimulus from the contemplation of time-honoured, unblemished memorials of those who have preceded them in the race and battle of life. But these truths are now universally admitted, and having been enforced by abler pens, it is unnecessary to insist further on them here.

Having in the former volume made some remarks on the origin and use of seals, it is needless to enlarge on the subject. To those who have studied Heraldry their value is well known; their importance, indeed, can hardly be overrated. In seals we find the trustworthy records of the progress of Heraldry from the earliest periods. It is truly said by a learned and graceful writer on the subject, in his allusion to Nisbet's estimate of the importance of seals, that "the worthy herald would unquestionably have been nearer the truth if he had asserted that they form the *most* authentic, as well as the earliest record of heraldic bearings."[1]

That the practice of Heraldry has not at all times been consistent with its principles must be freely admitted; and from the sixteenth century to a very recent period there are numerous examples of a wide departure from the simplicity and purity of an earlier age. Before the end of the seventeenth century, the learned Sir William Dugdale makes pathetic complaints respecting the confusion that prevailed in his time "through the liberty taken by divers mechanicks," whereby, he says, "the true use of arms will be utterly forgot." The experience of two centuries, however, has proved his fears in this respect to be groundless. With greater reason he laments the practice, then growing up, of differencing the "younger brethren," who establish new houses, with the "petty distinctions" of "crescents, mullets, martlets," etc., and strongly

[1] *Law and Practice of Heraldry in Scotland*, p. 189. By George Seton, Advocate. Edinburgh, 1863.

condemns the custom of crowding numerous quarters into one shield. In this we sympathize most cordially, considering that scarcely anything can be more subversive of that clearness and simplicity which constitute the first principles of heraldry. As a general rule, more than four quarters, that is, different coats, should not be admitted. We do not, of course, deny that several families are justly entitled to carry even as many as a hundred coats; but we quite agree with Sir William Dugdale, that "except it be to be made in a pedigree or descent to lock up in an evidence chest, thereby to show men's titles, or the alliances and kindreds of their houses, I see not to any use in the world they serve." Speaking generally, the object of quartering the arms of different families is to preserve from extinction those which have ended in an heiress; but in earlier times this object was sufficiently attained by combining the principal charges of each. Thus, for example, Stuart of Ochiltree surmounted the fess chequé with the bend and buckle of Bonkle on marrying the heiress of that family.

The great changes that have taken place in the habits of mankind, and the different objects of pursuit that now prevail in society, may, in a great measure, account for the neglect into which heraldry fell. But to us it does not appear that even in the present utilitarian and wealth getting age there is any insuperable obstacle to the practice of heraldry and genealogy in all their purity and usefulness, — nothing, indeed, to prevent their going hand in hand with the arts, manufactures, and commerce of the present time, as formerly they went hand in hand with the sterner occupations of war and conquest. That they will do so there seems every reason to believe, when such enlightened attention is directed to the subject. And when we find that the duties connected with the office of the Lyon King of Arms are now efficiently discharged, we may hopefully regard it as a guarantee that the "noble" science will not again be disgraced by the absurdities and "heraldic anomalies" which contributed in no small degree to expose it to ridicule.

When the former volume was published, it was therein stated that a

large additional number of seals was expected to be acquired from the
collections of the late General Hutton, which had just then been discovered. It was well known that General Hutton had for many years
been collecting materials, chiefly with the view, it is believed, of forming
a history of the monasteries and religious houses of Scotland. His
labours, however, were interrupted by his being called to the discharge of
his duties in Ireland, where he died soon after.

This collection was lost sight of for some years, but, in 1851, was
discovered through the accident of a fire in the house in which they
had been kept ; and by the liberality of the Rev. H. Hutton, son of the
General, the exertions of the late lamented Patrick Chalmers of Aldbar,
and of Mr. Albert Way, to whose intelligent zeal in archaeology this country
is so much indebted, the entire collection, comprising MS. notes and a
large number of casts from seals, were presented to the Advocates'
Library. This latter portion (the casts), with two MS. volumes, was subsequently presented to the Society of Antiquaries of Scotland.

The two MS. volumes are in 4to, bound in calf, and lettered "*Sigilla*" and
"*Seals*" respectively. The volume of "*Sigilla*" is nearly filled with drawings and notes of seals and charters, mostly by General Hutton himself, and
are exceedingly well executed ; many, unfortunately, are in an unfinished
state, but faithfully characteristic of the original. The volume named
"*Seals*" is only partially filled, and not nearly so interesting as the other,
though the sketches are equally good. These volumes recall to our recol-
lection a similar one in Dr. Rawlinson's collection, in the Bodleian Library,
to which we shall afterwards more particularly refer, and which we con-
sider General Hutton had seen and taken as a model : and the manner in
which it is so far carried out causes regret that it was not fully completed.
The collection of Casts has furnished several new and interesting seals,
which appear in the present volume ; but from the circumstance that there
are numerous duplicates of the same seal, and also that many others have
been obtained from other sources, the collections of General Hutton,
though apparently extensive, are not so in reality ; so that now their value

and interest, though by no means to be underrated, cannot be said to
realize the expectations of those who knew the care and accuracy with
which he made his researches. In the following pages reference is made
to this collection thus, " General Hutton's Collection ;" and where notes
and inscriptions occur in the MS. volumes, they are given within inverted
commas, along with the number of the page in the " *Sigilla*" or " *Seals*,"
as the case may be.

The collection of Dr. Rawlinson, in the Bodleian Library, is well known
to all who have given any attention to the subject of Seals. It consists of
several hundred original matrices, in excellent preservation, embracing,
however, very few Seals connected with Scotland or England,— Italy and
other continental countries supplying the chief examples ; and for the
illustration of the art as practised in those countries, the collection is
very valuable. The interest of it is further enhanced by numerous MS.
notes by Dr. Rawlinson himself, and especially by a MS. volume, folio,
containing drawings in pen and ink,— rather sketchy, but very effective ;
some engravings of seals (which appear to have been executed for the
purpose of illustrating " *The English Topographer*," written by Dr. Rawlinson) ; and an Introduction and Notes, in Italian, by the Abbé Valore,
in the year 1709. The Rev. Henry Coxe, the learned and obliging
Librarian of the Bodleian, is, we believe, arranging and compiling a Catalogue of this interesting collection. For the advantage we derived from
it we are indebted to Mr. Coxe, whose courtesy and attention we are
happy here to acknowledge.

Among the numerous evidences of the contest between this country
and her powerful neighbour, England, still preserved in H.M. Record
Office, a large and most interesting portion belongs to that unfortunate
period following the death of Alexander III., when the ambitious designs
of Edward I. were prosecuted with an energy and sagacity well cal-
culated to insure success, but which ultimately proved futile. In this
portion are found many of the Deeds of Homage which the power
of the unscrupulous King extorted from a distracted people. They

are mostly drawn up in Norman-French, then in use for all legal instruments, in a most correct and cautious form of phrase, and attested by
appending the seal of each one giving the homage. The Author of the
present Catalogue, at the expense of certain noble and generous patrons
of heraldic studies, went to London and made a careful examination
of all the seals remaining, and had photographs taken of the greater number, of which the negatives are preserved, but from bad light and other
causes they are not in general very satisfactory.

    Not content with receiving the homage of the barons and magnates
of the land, Edward insisted also on the commonalty, or inhabitants
of burghs, counties, or districts, giving homage collectively, with the
attestation of their respective seals. Hence we find deeds containing
between one and two hundred names of the inhabitants of certain
districts, comprising a number of most extraordinary ones, and offering a rich field for ingenious speculation to any who are fond of the
study of surnames. Among them are many still known and common,
after the lapse of nearly six centuries, while others have quite disappeared.
It is possible, indeed highly probable, that some of these homages were
taken from corporate bodies,—one, in particular, now unfortunately
almost illegible, to which are still appended nearly a hundred seals :
while many have evidently dropped off. These will be found described
in the following pages, under the initial letter of the names – the first
instance occurring at No. 48, where, by an unaccountable and unpardon-
able oversight, the date is given as 1292 instead of 1295-6.

    Some of these deeds are still in good condition, and are excellent
specimens of the caligraphy of the age : others are more or less injured,
while of many nothing remains but the seal and tag by which they were
appended. Most of these detached or loose seals will be found described
in a distinct class in the Appendix, which contains also some interesting
seals that have been met with during the progress of the present work
through the press, and which could not be inserted in their proper places.
In the Appendix will also be found some curious and remarkable seals.

which, notwithstanding repeated attempts, we are quite unable to identify. They are here described as accurately as possible, in the hope that some satisfactory explanation may yet be given.

It may perhaps be objected to this work, professing to be devoted to *Scottish* Seals, that there are some decidedly English names, which should have no place in this collection; but, in reply to this, it may be stated that such instances are not numerous, and in no case, we believe, do they occur but where they are more or less directly connected with Scotland, such, for example, as donations to religious houses in Scotland, in the earlier periods, by the English families of De Vesci, Vipont or Vetripont, and others; while in the subsequent periods, there can be no doubt that many English names were introduced which are now claimed, rightfully or wrongfully, as of this country.

The Ecclesiastical and Municipal Seals, as well as those of other corporate bodies, are far from being devoid of interest to the archæologist or herald, though to the latter certainly not equal in importance to Family or Baronial Seals. This may in a great measure be ascribed to the fact that the practice of Corporations bearing Armorial Ensigns is of comparatively modern date, and also that the designs assumed by the burghs as their Armorial Ensigns are, in most instances, quite incongruous with true heraldry.

In regard to the Armorial Ensigns of the different dioceses of Scotland, there can be no question of their very late assumption; and only five of them have had their arms recorded in the Register of the Lord Lyon, on the establishment of that record in 1672, and thus obtained due authority for carrying them.

The seals of the Bishops, particularly those of an early period, are rich and beautiful in design, the paternal shield being fitly introduced; but they are only personal, and cannot be regarded as belonging distinctively to the see.

With respect to the Armorial Ensigns of the burghs of Scotland, they are, in most instances, merely a transference of the design of the earlier

c

common seal to a shield, thereby giving the desired heraldic character. All the burghs, royal or baronial, had, from the earliest period of their erection, a common seal, the design on which was generally the patron saint, or the shield of the baron from whom its privileges were held; it is not, however, at all common to find any shields on the early burgh seals. Aberdeen is the earliest instance we have yet met of proper armorial ensigns on a common seal. The date is 1430, and the original matrix—a fine specimen of the art—is still preserved; it is fully described in the "*Descriptive Catalogue of Ancient Scottish Seals*" (No. 1146-7), and is there said to be in the possession of the Corporation of Aberdeen; but this is found to be an erroneous statement, which we gladly take the present opportunity of correcting. It is in the possession of Mr. William Smith of Springbank, Aberdeen, who accidentally met with it among a lot of old metal exposed for sale in a broker's shop.

The shield has for supporters two lions, not leopards, which, for some time past, have appeared as such. This will recall to the archæologist the similar, though converse, change in the Royal lions of England, which in early times were blazoned leopards. There was certainly some fanciful distinction given to the names to indicate the attitude of the animal. Thus, when it was passant, gardant, it was called a leopard, and when rampant, a lion; but these absurd terms have happily long since passed away. It should be observed that the shield bears only one triple-towered castle within the Royal tressure, though very soon after this date (A.D. 1444) we meet with the three as they are now carried (see No. 1148, "*Descriptive Catalogue of Ancient Scottish Seals*"). It is believed that these seals are the only existing evidence of the armorial ensigns of Aberdeen till a much later period. That they were, as it is said, granted by King Robert Bruce, A.D. 1319, is very probable, and they are recorded in the Register of the Lord Lyon in 1672.

Neither on the seals, nor in the Patent, is any crest given, and we believe the burgh has never assumed any. The Patent recording the arms of Aberdeen is printed verbatim in the Appendix to Mr. Seton's work ("*The Law and Practice of Heraldry*"), and is, in fact, merely a description of the ancient seal above referred to. As regards one side, however, the Patent is most unaccountably erroneous. It describes it as representing St. Michael and three children in a boiling caldron, while, in truth, it is St. Nicholas, the patron saint, restoring to life the three children killed and hidden in a barrel by an innkeeper at Myra. Such a mistake betrays either lamentable ignorance or gross negligence.

Edinburgh, the capital of the kingdom, does not appear to have assumed proper armorial ensigns so early as some other burghs of less importance. It was only in the year 1732 that the city obtained a Patent for the armorial ensigns now borne, though they had certainly been carried several years previously, as we find them, among other instances, appearing on a pack of heraldic playing cards, bearing the date 1691, now in the possession of Mr. David Laing. They are a very interesting series, and well executed. For a fuller account the reader is referred to "*The Law and Practice of Heraldry,*" p. 211, note. In these cards the crest is a castle; the anchor, with cable, having been subsequently adopted. The armorial ensigns of the capital are too well known to require particular description here, but we may be allowed to observe that the castle on the rock is derived from the early common seal used in the fourteenth century, and on a different seal of the sixteenth century the same type appears (see Nos. 1156-7, in "*Descriptive Catalogue of Ancient Scottish Seals*"). The counter seals of both exhibit the patron saint (St. Giles), in the latter instance accompanied with his fawn. It may not, perhaps, be too violent an assumption to suppose that these furnish the type of the present graceful supporters, to account for which various fanciful theories have been propounded. Bearing in mind the distractions of the seventeenth century, and the almost total neglect of Art and Heraldry consequent thereon, it does not appear to us at all extravagant to suppose that the patron saint has undergone a transition; and as his symbol, the fawn or hart, could not, we may suppose, be regarded as

savouring of superstition or idolatry, it was therefore retained, without
any attempt to adapt it to the altered feeling of the times.

At a Convention of the Royal Burghs, held in A.D. 1673, a resolution
was passed, recommending such of the royal burghs as had not obtained
armorial ensigns immediately to apply to the Lord Lyon for a grant, but
very few seem to have complied with this proper recommendation; for it
appears that of all the burghs in Scotland using arms, only nineteen are
recorded in the Register of the Lord Lyon, and any of the others carrying
heraldic insignia are doing so without legal authority.

The practice of corporations, whether for municipal or for trading
purposes, having armorial ensigns, has prevailed for a long period; and
from the present rapid increase of various trading companies (limited)
seems likely to prevail to an *unlimited* extent. In such cases the
existence of the King at Arms seems quite ignored, and every banking
or other company appears to feel justified in assuming any heraldic
blazon it thinks proper. It should, however, be known that the right
of any corporate body to use armorial ensigns rests entirely on the
same grounds, and is given and protected by the same constitutional
authority, as that of private individuals and families; and no corpora
tion, municipal, ecclesiastical, or commercial, can *legally* use them with
out such authority.

The only portion of the author's duty now remaining to be discharged
is to acknowledge the kind assistance of his supporters. Words can but
feebly express the gratitude he feels towards those old and tried friends
who, through seasons of both shade and sunshine, have ever been ready to
encourage and assist his humble labours. By their benevolent exertions,
he is now enabled to pass the evening of an active, and, he trusts, not alto-
gether useless life, in comparative rest and quietness, which he intends to
devote, so far as time and health permit, to the pursuit of the same
objects in which the best part of his life has been spent.

While thus rendering thanks to his friends generally, the author feels

bound to make particular reference to the assistance afforded in the pre-
paration of the present work by Lord Lindsay, Mr. Cosmo Innes, Mr.
Mark Napier, Mr. David Laing, Mr. Joseph Robertson, Mr. William Fraser,
Mr. George Burnett (the present able and energetic Lyon-depute), Mr.
Albert Way, Rev. W. Greenwell, Durham, etc.; but especially should
he mention Mr. George Seton, to whom personally, as well as to his excellent work, "*The Law and Practice of Heraldry in Scotland*," the author
is indebted for much valuable information. His thanks are also due to
Mr. Ebenezer Colquhoun, of Glasgow, from whose extensive and rich
collection he has obtained many of the burgh seals contained in this
volume.

To the authorities directing the National Institutions of the country,
as H.M. Record Office, the General Register House, the Colleges of St.
Andrews, Aberdeen, and many others,—his thanks are due for the courteous
and liberal manner in which they allowed access to their invaluable collections. He has also to acknowledge the same liberality and courtesy from the
Venerable the Dean and Chapter of Durham, whose collection of charters
and other documents of an early period, referring to Scotland, is, for
antiquity, extent, and interest, unequalled, ample testimony of which will
be found in the following pages.

To Mr. James T. Gibson Craig the author is indebted for the liberal
contribution of twelve woodcuts; and to Sir William Stirling Maxwell,
Bart., M.P., for the loan of those illustrating the Stirling seals. He is under
the same obligation to Mr. Innes for those pertaining to the Inneses; to
Sir John Lawson, Bart., of Brough Hall, for the loan of No. 134; to
Dr. Bedford, House-Governor of Heriot's Hospital, for No. 1194; to the
Archaeological Institute for Nos. 426, 663, 966, 967; to the Society
of Antiquaries of Scotland for No. 1294.

In conclusion, the author begs to remark that, in works of this kind,
it is almost impossible to escape falling into some inaccuracies, and,
accordingly, he is fully prepared to hear of mistakes which he would
gladly have avoided. He is conscious of sparing no pains or care, and he

hopes his readers will be lenient to his errors, and accord the same favour to this as to the former volume.

In justice to the artists employed in the illustrations, it should be mentioned that most of the woodcuts are the work of Mr. Corner, while a few are by Mr. Adam. Some of the lithographic plates are executed by the well-known Frederick Schenck, who was prevented by circumstances from completing the whole. The remainder, however, have been finished by Mr. Ritchie in an equally creditable manner. The frontispiece is the production of Mr. Dick, from drawings by Mr. Wallace, a young artist of promise, who also furnished drawings for some of the other illustrations. All are executed by Edinburgh artists, and it is hoped these specimens of their skill will be no discredit to the Schools of Art in the metropolis of Scotland.

H. LAING.

ELDER STREET,
EDINBURGH. *May* 1866.

CONTENTS.

ERRATA.

No. 285, p. 49, *for* ninth *read* eleventh.
1042. p. 176, *read* Plate XI. fig. 1.
1105, p. 191, *read* Plate XI. fig. 8.
1155, p. 203, *read* Plate XV. fig. 4

LIST OF THE PLATES.

SUPPLEMENTAL

DESCRIPTIVE CATALOGUE.

ROYAL SEALS OF SCOTLAND.

1. MATILDA. Daughter of Malcolm Can-More and Queen of Henry I. a.d. 1100, died 1118. *Plate* I. *Frontispiece, fig.* 5.

A remarkably fine seal, oval shape, a full length front figure of the Queen, holding in her right hand a sceptre ornamented with a dove, in her left the orb and cross. The mantle is disposed in elegant folds from the shoulders to the feet.

" SIGILLUM MATHILDIS . . . . . . AN . . . . . . EI GRACIA REGINAE ANGLIE."—*Appended to a Grant of Carham on the Tweed to St. Cuthbert's, Durham.—Dean and Chapter of Durham.*

2. NORTHUMBERLAND, PRINCE HENRY, EARL OF, Son of David I. *Plate* IV. *fig.* 1. A Knight on horseback to sinister, in ring mail apparently, conical helmet without nasal, sword in his right hand, and on his left arm a shield quite plain.

" SIGILLUM HENRICI COMITIS NORTHUMBERLANDIE FILII REGIS SCOCIE."—*Appended to the Charter of Edenham and Nisbet to St. Cuthbert at Durham.—Dean and Chapter of Durham.*

3. ALEXANDER III. a.d. 1249-85. Great Seal. *Plate* II. *fig.* 1.

Unfortunately this is but a fragment; the part remaining is, however, very perfect, and proves it to have been a fine seal.

The King on horseback galloping to the sinister, in chain mail with surcoat, his right hand extended holding a drawn sword, his left bearing a shield charged with the Lion Rampant and Royal Tressure, which are also repeated on the

A

fore and back parts of the housings or caparisons, here first introduced on the
Royal Seals; a small saddle-cloth only having been used previously. Not a
letter of the inscription remains.

4. COUNTER SEAL OF THE LAST. *Plate* II. *fig.* 2.

> The King sitting on a stool, robed; the cloak, lined with ermine, thrown over
> his shoulders, and fastened on the breast; both arms extended, his right hold-
> ing a sword. The front of the stool is ornamented with two lions' heads, full
> faced, enclosed within cusped panels.— *Appended to a Charter by the King
> confirming the gift made by the late John de Montfort to his brother, Alex-
> ander de Montfort, of the lands of Elstaneford, etc., etc. in Haddington.
> At Kincardine,* 29th *August* (" *third year of our reign*"), A.D. 1252.—
> *D. Laing, Esq.*

> This is an interesting and exceedingly rare seal. The one here described, and one
> recently discovered among the Burgh Charters of Inverness, are the only instances
> known. Until their discovery it was supposed that only one Great Seal—the
> fine one so well known—had been used during the reign of Alexander III.
> This example, however, proves the contrary beyond a doubt. It was probably
> discontinued when the King attained his majority and assumed the sole govern-
> ment, about A.D. 1262.

5. EDWARD I. OF ENGLAND.

> A fine seal, rather less in size than the Great Seals of the period, but of similar design.
> The King is seated on a throne, his feet resting on two lions; his right hand,
> resting on his knee, holds a sceptre supporting a crown.

> " SIGILLUM EDWARDI DEI GRACIA R[EGIS ANGLIE] DNI HIBERNIE." This inscription is
> completed on the Counter Seal.

6. COUNTER SEAL OF THE LAST.

> A shield bearing three lions passant gardant—England.

> " ET DUCIS AQUITANIE AD REGIMEN REGNI SCOCIE DEPUTATUM."—*From the Collection
> of Casts taken by the late Mr. Doubleday, who calls it the seal of* " EDWARD I.,
> DEPUTY-GOVERNOR OF SCOTLAND," *but gives no date.*

7. JOHN BALLIOL. *Plate* II. *fig.* 3.

> Merely a fragment of a fine large seal. The King seated on a throne, the arms and
> legs of which are formed of the heads and feet of animals; on the robe of
> the King is embroidered the Lion and Tressure of Scotland. All that remains

of the inscription is " . . . . EI GRACIA REG. . . ."—*From the same Collection as the last.*

To this seal Mr. Doubleday gives the name of "JOHN SOULY, CUSTOS REGNO SCOTIE 1301;" but there can be little doubt it is the seal adopted and used by the national party in Scotland after the surrender of John Balliol in 1296, when the Great Seal used by that sovereign was withdrawn; his authority, however, was still recognised in Scotland.

8. MARGARET (LOGIE), QUEEN OF DAVID II.

A fine seal, but much injured. A full-length figure of the Queen. Above is a shield bearing Scotland, and supported by two Lions. At each side is a shield, the dexter bearing a fess chequé (Stuart), and the sinister three bars wavy for Logie; only a few letters of the inscription now remain.

"SIGI . . . . RE . . . ."—*Appended to Acquittance of the Queen's rights to certain duties, 23d June 1372.—H.M. Record Office.*

9. JAMES I. A.D. 1406-1436.

The Quarter Seal. The design on this and the following is the same as on the upper half of the Great Seal.—*Vide No. 41 of "Descriptive Catalogue of Scottish Seals."*

"JACOBUS DEI G SCOTTORUM."

10. COUNTER SEAL OF THE LAST.

"JACOBUS DEI G. SCOTTORUM."—*Appended to a copy of an Act of Parliament, 26th May 1424.—Dean and Chapter of Durham.*

Some interesting notes regarding Quarter Seals, their uses, etc., by Joseph Robertson, Esq., will be found in *"Proceedings of the Society of Antiquaries of Scotland,"* vol. ii. part iii. p. 428.

These Quarter Seals seem peculiar to Scotland, and are now scarcely known beyond the Border, though at an early period used in England; but since the time of Edward I. they have disappeared. The impressions here described, and still remaining, appended to Scottish instruments, are ample proof of their continued use in this country. On the accession of Her present Majesty a new Quarter Seal was executed, as well as a new Great Seal, and both are now in the custody of the Keeper, the Earl of Selkirk, in H.M. General Register House. The original silver matrix of the Quarter Seal of George IV. is in the Museum of the Society of Antiquaries of Scotland, having been presented, together with the Great Seal of George III., to them by H.M.

CATALOGUE OF

Privy Council, through the Right Honourable Earl Granville, the President, in 1857.

11. **JOAN BEAUFORT**, Daughter of John Earl of Somerset, Queen of James I. Married A.D. 1424; died A.D. 1445. *Plate III. fig. 3.*

This has evidently been a fine seal, but is now very imperfect. The shield is a lozenge, and the only example of that shape yet met with among Scottish seals. It bears Scotland, impaling the Queen's paternal arms; France and England quarterly within a bordure componé, in reference to the illegitimacy of the Queen's father. See page 208 of Seton's " *Law and Practice of Heraldry in Scotland* " (the best work on the subject yet published). The dexter supporter alone remains,—apparently a unicorn, and if so, this is probably the earliest example of that animal appearing as a supporter.

" [JOANNE DEI GRA] REGINE SCOTTORUM."—*Appended to Agreement made at Stirling between Joan, Queen of Scotland, and Sir Alexander Livingston and others, regarding the Custody of the young King James II., 4th Sept. 1439.- H.M. General Register House.*

12. **JAMES IV.** A.D. 1488-1513. *Plate III. fig. 1.*

The Quarter Seal. The description of No. 9 applies equally to this; *vide No. 51 of " Descriptive Catalogue of Scottish Seals."*

" JACOBUS DEI G SCOTTORUM."

13. **COUNTER SEAL** of the last. *Plate III. fig. 2.*

" JACOBUS DEI G SCOTTORUM."—*Appended to a Commission by the King, 26th May 1503.—H.M. Record Office.*

14. **MARGARET TUDOR**, Daughter of Henry VIII., and Queen of James IV. Married A.D. 1503; died 1542.

The Queen, crowned and sitting on the ground, fondling a unicorn in her lap; in the background are sunflowers and two kids. *From a letter of the Queen to Henry VIII., dated 11th April 1513.—British Museum*

15. **MARGARET TUDOR.**

Merely a shield, bearing Scotland impaling France and England quarterly. Above the shield is a crown, with foliage at the sides.—*From the Collection of Casts taken by the late Mr. Doubleday.*

16. **MARY.** A.D. 1542-1567

The Quarter Seal, but a mere fragment. The preceding description of No. 9

applies equally to this and the following one of James VI.; *vide No. 58 of "Descriptive Catalogue of Scottish Seals."—Appended to a Precept of Sasine for Thomas Scott in the lands of Over Collane, by King Henry and Queen Mary, 6th February 1564-5.—H.M. General Register House.*

17. JAMES VI. A.D. 1567-1625.

The Quarter Seal, as before; *vide No. 67 of "Descriptive Catalogue of Scottish Seals."—Appended to a Precept for Infeftment in favour of John Graden in a husband-land of Graystoneig Wester, co. Berwick, 26th December 1610.—H.M. General Register House.*

18. JAMES VI. A.D. 1567-1625.

A Privy Seal. The arms of Scotland beneath an arched crown, from the sides of which issue a scroll with the words, "IN DEFENS." The collar of the Order of the Thistle surrounds the shield.—*British Museum.*

19. CHARLES PRINCE OF SCOTLAND AND WALES, AFTERWARDS CHARLES I. Born A.D. 1600; succeeded his father, 1625; beheaded 1648-9. *Plate II. fig. 4.*

An armed knight on horseback galloping to the dexter, a drawn sword in his right hand; his left, without the usual protection of the shield, holding the reins. A plume of feathers decorates the helmet. The housings of the horse are richly ornamented, having a thistle on the fore part, and the arms of Scotland on the hind. The background presents the view of a city and landscape. There is no inscription.

20. COUNTER SEAL OF THE LAST. *Plate II. fig. 5.*

Quarterly, first and fourth, Scotland. Second quarterly, France and England. Third, Ireland. Above the shield a crown of five points. Supporters: Dexter, a unicorn gorged and chained; sinister, a lion.

"MAGNUM SIGILLUM CAROLI SCOTIÆ WALLIÆ PRINCIPIS ROTHESAIÆ DUCIS, ETC." *Communicated by the late John Mackinlay, Esq.*

21. COMMONWEALTH. A.D. 1649-1653.

An oval-shaped shield, with scroll ornaments. Quarterly, first and fourth, the Cross of St. George, for England. Second, a saltire (St. Andrew's Cross), for Scotland. Third, a harp, for Ireland. The background is diapered with thistles and saltires.

"THE PRIVIE SEALE FOR SCOTLAND."—*From the Collection of Casts taken by the late Mr. Doubleday.*

22. COMMONWEALTH. A.D. 1649–1653.

Oval shape. A lion sejant supporting a shield quarterly as in the last.

"PAX QUÆRITUR BELLO."—*Appended to Commission appointing Sir Archibald Johnston of Warriston to be one of the Commissioners of the Treasury and Exchequer in Scotland, 20th September* 1657.—*H.M. General Register House.*

23. STUART, HENRY BENEDICT, SECOND SON OF JAMES FRANCIS EDWARD, "THE CHEVALIER ST. GEORGE," AND GRANDSON OF KING JAMES II. OF ENGLAND. Born A.D. 1725; elected Cardinal A.D. 1747; died 1807; the last of the royal line of the Stuarts, and sometimes designated Henry IX.

A square-shaped shield, bearing, first and fourth, France and England quarterly: second, Scotland; third, Ireland,—over all a crescent for difference, as second son. Above the shield a crown of five points of fleur-de-lis and crosses patté, and over it a cardinal's hat, the tassels falling on each side of the shield.

This pretty and interesting seal is affixed to a license or permission by the Cardinal to shoot on some lands belonging to him.

SEALS

BARONS AND MAGNATES, ETC., OF SCOTLAND.

---

24. ABERCROMBY, GEORGE, OF PITMEDAN.

A chevron between three boars' heads erased.

"S' GEORGI ABERCROMBIE."—*Appended to a Charter by G. Abercrombie of the lands of Pitmethane to his son James Abercrombie, and Mariota Hay his spouse. 13th July 1537.—Forglen Charters.*

25. ABERNETHY, ALEXANDER.

Quarterly, first and fourth, a lion rampant, debruised with a ribbon, for Abernethy. Second and third, three passion nails conjoined in base (doubtless meant for three piles), for Wishart of Brechin; in middle chief point, a crescent as a difference.

"S' ALEXANDRI ABIRNETHI."—A.D. 1491.—*Communicated by Lord Lindsay.*

26. ABERNETHY, WILLIAM, OF BIRNY.

Quarterly, first and fourth, Abernethy, as in the last. Second and third, three piles.

"S' WILLELMI ABERNETHIE." Detached seal, on the label of which is written. "W. Abernethy's seall. 1st July 1595."— *Hutton's "Sigilla," p. 89.—Library of S. A. Scot.*

27. ADAMSON, JOHN.

A mullet between three cross-crosslets fitché.

"SIGILL JONIS ADAMSON."—*Appended to a Charter by John Adamson to John Scott, 28th July 1529.—Edinburgh Charters.*

28. AGMONDESHAM, WALTER.

Oval shape. A female head, not on a shield.

"S' WALTERI D' AGMODESHAM."—*Appended to a Fragment of a Document, probably of the twelfth, or early in the thirteenth century.—H.M. Record Office.*

29. AINSLIE, JAMES, of Darnick, Burgess of Edinburgh.
    A cross flory, voided in the centre, between three mullets and a crescent, the latter
        in the dexter base. A rose and thistle foliage at the top and sides of the shield.
    "s' JACOBI AINSLIE."—*Appended to a Precept of Clare Constat by James Ainslie,
        Merchant and Burgess of Edinburgh, in favour of Andrew Thomling, of some
        land at Darnick, 8th August 1617.—Dr. J. A. Smith, Secy. S.A. Scot.*

30. AITCHISON, MARK, Portioner of Ballencreif.
    An eagle displayed.
    "s' MARCI ACHISOUN."—*Appended to a Charter of Rowieston's lands in Ballen-
        crieff to John Murray of Eddilston, 7th June 1606. Ballencreiff Charters.—
        Lord Elibank.*

    ALBANY, DUKE OF. *Vide* Stuart.

31. ALDCAMBUS, EDWARD.
    A fleur-de-lis, not on a shield.
    "SIGILL. EDUARTDI DE ALDECAMBUS," the two last letters are in the field of the seal.
        —*Appended to a Covenant between B., the Prior, and Convent of St. Cuthbert's,
        Durham, and Edward de Aldecambus, regarding the lands of Aldecambus and
        Lumsden, A.D. 1198.—Dean and Chapter of Durham.*

32. ALDENGRAY, GREGORY MARSHAL of.
    A flower, not on a shield.
    "s' GREGORII MARSCAL."—*Appended to a Charter of " Gregori filii Marescal de
        Aldengrey" of sixty acres of land of Birkensyde in Aldengrey to the Convents
        of St. Ebbe, Coldingham, and St. Cuthbert's, Durham, A.D. 1255.—Dean and
        Chapter of Durham.*

33. ALDENGRAY, MICHAEL.
    A fleur-de-lis, not on a shield.
    "SIGILL. MICHAELL FILII EDWARDI."—*Appended to a Charter of some lands at
        Aldengrey to the Convent of St. Ebbe at Coldingham.—Dean and Chapter of
        Durham.*

34. ALDENGRAY, THOMAS.
    A cross potent between four stars of eight points, and surmounted by one of the same,
        not on a shield.
    "s' THOME FILII ROBERTI."—*Appended to a Charter by Thomas, son of Robert*

son of *Mathew de Aldengrey to St. Cuthbert and St. Ebbe, and the Prior and Monks of Coldingham, of twenty acres of land and some messuages at Aldengrey,* A.D. 1275.—*Dean and Chapter of Durham.*

35. ALEXANDER, Son of William.

A shield bearing a fess between three rams' heads couped. Above the shield a demi figure of a lady, and at each side of it a lion couchant. All in centre of pointed tracery.

"S' ALEXDR FIL. WILELMI."—*Appended to Charter by John Crab, Burgess of Aberdeen, to the Priory of the Carmelites, of ten marks yearly, 24th August* 1382.—*Marischal College, Aberdeen.*

36. ALMAN, ROBERT.

Three piles in point, surmounted by a bend, above the shield a cinquefoil between an estoile on the dexter and a crescent on the sinister, at each side of the shield, two mullets of six points.

"S' ROBERTI D'ALMAN."—*Appended to "Demission" of a Toft, in the "Villa de Coldingham," to William de Howburn,* A.D. 1304.—*Dean and Chapter of Durham.*

37. ALMAN, WILLIAM, of Durham.

On a chevron between three leopards' heads, jessant fleur de-lis, three mascles. The shield surrounded by tracery.

"SIGILLUM WILLELMI ALMAN."—*Appended to a Charter by William Alman and Margaret his wife, of some lands at Durham, to St. Cuthbert's of Durham,* A.D. 1355.—*Dean and Chapter of Durham.*

38. ANGUS, HUGONIS.

A Galley.

"S' HUGONIS ANGUS."—*Appended to a Charter by Hugh Angus to William Buisse, of an annual rent of two shillings from ground in Driemesdale Street,* A.D. 1456.—*Inverness Charters.*

39. ANGUS, WILLIAM.

A saltire couped between the letters V-A; in chief a mullet.

"S' DNS WILELMUS ANGUS."—*Appended to Charter by John Burne, Burgess of Dunfermline, and Janet Michel, his Spouse, of the eighth part lands of the Hospital of St. Leonard's, to Henry Trumble, 4th November* 1607.—*Elgin Charters.*

ANGUS, EARL OF. *Vide* DOUGLAS and STUART.

B

40. ANNAND, DAVID.

A saltire and chief, with a label of three points.

" SIG. DAVIDIS ANNAND FILII DAVIDIS ANAND MILITIS.   *From Sir Jas. Balfour's
Collection."—Hutton's " Sigilla," p. 167.

41. ANNAND, JOHN.

A saltire and chief.

" SIGILLUM JOANNIS DE ANNAND. — 20 Nov* 1421."—Hutton's " Sigilla," p. 154.

42. ANNAND, WILLIAM.

A boar's head, not on a shield.

" s' WILLMI [D]E ANANT."   Detached Seal.—H.M. Record Office.

43. ARBEKE, THOMAS.

A boar's head couped, contourné in base, with a garb (?) and a mullet in chief.

" s' THOME DE ARBEKE."—Appended to an Instrument dated 1453.—G. Smythe,
Esq. of Methven.

44. ARBUTHNOT, SIR ROBERT—" DE EODEM DMI DE PORTARISTOUN."

Much injured.   Conché: a crescent between three mullets.   Crest on a helmet, a
peacock's head.   The sinister supporter, a dragon without wings, alone remains.
The dexter supporter was probably the same.

" s' ROBERTI [ARBUTHNOT]"—Appended to a Precept of Sasine of the lands of
Orchartoun, in favour of John Wishart, A.D. 1493.   Communicated by W.
Fraser, Esq.

45. ARBUTHNOT, ROBERT, OF BANFF.

On a fess between three mullets, a crescent.

" s' ROBERTI ARBUTHNOT."—Appended to the same Instrument as the last.

46. ARDROSSAN, GODFREY.

An antique gem, very indistinct.

" SIGILL. SECRETI."—Appended to the Homage of Godfrey de Ardrossan, A.D. 1295.
—H.M. Record Office.

ARGYLE, EARL OF.   Vide CAMPBELL.

47. ARISTOTLE, GILBERT.

Oval shape.   A well-executed device of a monk sitting before a lectern.

" s' GILBERTI ARISTOTELIS."—Appended to a Letter of G. Aristotle, acknowledging *dependence of the Church of Brankston to the Prior and Monks of Durham, about A.D. 1230.—Dean and Chapter of Durham.*

48. ARLESAY, WILLIAM.

A curious device; a lion coiled in front of a tree, on the dexter side of which is a hare playing on a tambourine, and on the sinister a fox (?) playing on pipes.

" s' WILLI DE ARLESAY."—*Appended to Homage Deed, supposed to be that of some corporate body, A.D. 1292.—H.M. Record Office.*

49. ARNCAPLE, MARY.

A device of a stag's head cabossed; between the antlers a dog and a fleur-de-lis.

" s' MARIE DE ARNCAPEL." *Detached Seal.—H.M. Record Office.*

50. ARNNER, ROGER.

A hunting horn, not on a shield.

" s' ROGERI DE ARNNER " *Detached Seal.—H.M. Record Office.*

ARRAN, EARL OF. *Vide* HAMILTON.

51. ASCOLE, OR DE ASCOLE, ROLAND.

A right hand apaumé, not on a shield.

" s' ROLAND DE ASCOLE " *Detached Seal.—H.M. Record Office.*

52. ASCOLOG, HECTOR.

On a shield, surrounded by tracery, two lions passant in pale.

" SIGILL. HECTORIS ASCOLOG."— *H.M. Record Office, London.*

ATHOL, EARL OF. *Vide* STUART.

53. ATKINSON, THOMAS.

On a chevron, three buckles.

" s' THOMAE FILIUS ADE."—*Appended to an Indenture, A.D. 1429.—Dean and Chapter of Durham.*

54. AUDREY, ROGER.

A chalice on a shield.

" SIGILLUM ROGERI DE AUDREY."—*Appended to a Charter by Roger Audrey and Wimart, his wife, of the tithes of the Mill of Ancroft, etc. etc. to the Church of Norhamshire and Islandshire.—Dean and Chapter of Durham.*

55. AULMAULDEN (?), WILLIAM.

    A device of a dog sitting up.

    " s' will. aulmaulden."—*Appended to the same Homage as No. 48, A.D.* 1292.—
    H.M. Record Office.

56. AURGOT, PETER.

    A stag's head caboshed, supported by two dogs, not on a shield; a mullet in the
    background, and a cross crosslet between the antlers.

    " s' petri de aurgot."   Detached Seal.—*H.M. Record Office.*

57. AVESANS, JOHN.

    A wheel ornament.

    " s' johis de avesans."—*Appended to the same Homage as No.* 48, A.D. 1292.

58. AYTON, HELIAS, of Upper Ayton or Eiton.

    Oval shape; an eagle displayed, not on a shield.

    " sigill. helie d[e e]yver eiton."—*Appended to a Charter by Helie Ayton, of
    lands at Upper Ayton, to St. Cuthbert's at Durham.- Dean and Chapter of
    Durham.*

59. AYTON, SIBILLA.

    An eight-leaved flower, not on a shield.

    " si. sibilla de aiton."—*Appended to a Quitclaim of all her lands in Upper Ayton
    to St. Ebbe's at Coldingham.—Dean and Chapter of Durham.*

60. AYTON, ADAM.

    A fleur-de-lis, not on a shield.

    " s' ade de aiton."—*Appended to a Charter by Ade, son of William, of some
    lands at Ayton, to the Priory of Coldingham, for building a mill.—Dean and
    Chapter of Durham.*

61. AYTON, ALICIA.

    Oval shape; a crescent between two mullets, not on a shield.

    " s' alicia ux patrici."—*Appended to a Charter by Alicia.* " quondam uxor
    Roberti filii Maurici de Ayton," *of some lands at Upper Ayton, to St. Ebbe's
    at Coldingham, A.D.* 1276.—*Dean and Chapter of Durham.*

62. AYTON, JOHN.

A front male head, not on a shield.

"s' JOANNES AYTON."—*Appended to a Grant or Special License by John, son of William de Ayton, to the Prior and Monks of St. Cuthbert at Durham and St. Ebbe at Coldingham, to make a Water-course through his grounds of Castelhull. A.D. 1327.—Dean and Chapter of Durham.*

63. AYTON, WILLIAM.

A hare or rabbit sitting, not on a shield, the seal itself being of that shape.

"s' WILELMI DE AYTON." Detached Seal.—*H.M. Record Office.*

64. BAIRD, ANDREW, OF LAVEROCK-LAW, AFTERWARDS OF AUCHMEDEN. Died A.D. 1543.

A fess between three mullets in chief, and a sanglier passant in base.

"s' ANDREE BAIRD."—*This and the two following are from sketches on the title-page of MS. "History of the Family of Baird, A.D. 1700," in the Advocates' Library, Edinburgh, which states that they are copied from leaden impressions, but does not specify any document to which impressions in wax are appended. The dates here given are from the same authority.*

65. BAIRD, GEORGE, OF ARDINGHAM. Died A.D. 1557.

On a fess three mullets, in base a sanglier passant, foliage at the sides and top of the shield; rather rude workmanship.

"s' GEORGII BAIRD."

66. BAIRD, WALTER, OF ARDINGHAM, SON OF THE ABOVE GEORGE BAIRD. Died A.D. 1599.

A fess between three mullets in chief, and a sanglier in base.

"s' WALTERI BAIRD."

67. BALCASKY, THOMAS, BURGESS OF PEEBLES.

A shield bearing a fess; in base a mullet, and the chief fretty apparently.

"s' THOME BALCASKY."—*Appended to an Indenture between William Dudingston of Southouse, and Thomas Balcasky, 13th August 1508. Blacktarong Charters.—Lord Elibank.*

68. BALLIOL, EUSTACE DE. *Plate IV. fig. 5.*

A fine seal, exhibiting a knight on horseback in chain mail, a sword in his right hand, and on his left a shield, charged with a carbuncle or escarbuncle.

"SIGILLUM EUSTACHIE DE BALLOLIO."—*Appended to an Agreement between Eustace*

Balliol, with consent of Hugh, his heir, the Monks of St. Cuthbert at Dur-
ham, and the Monks of St. Alban's, regarding the Church of Bywell, about
A.D. 1190.—Dean and Chapter of Durham.

69. BALLIOL, HUGH, Son of the above Eustace.   Died before 1228. *Plate IV. fig. 6.*
Similar equestrian design to the last, but in this the shield bears the orle with umbo
in centre.
"SIGILLUM HUGONIS DE BAILLIOLO."—*Appended to a Charter of some lands at*
Bacworth to " Radulf filius Aidani de Hindelic," early in the thirteenth century.
—Dean and Chapter of Durham.

70. BALLIOL, HUGH. The same person as the last.
A heart or pear-shaped shield, bearing an orle.
SIGIL. [HV]GONIS BA[LLI]OLO."—*Appended to Confirmation of the Church of Bywell*
to the Monks of St. Cuthbert at Durham.—Dean and Chapter of Durham.

71. BALLIOL, DERVORGILLA. Daughter of Alan Lord of Galloway, and Margaret, Daughter of David Earl of Huntingdon, Wife of John Balliol.
Plate V. fig. 1.
An elegant and well-executed design, on pointed oval-shaped seal.   A full-length
front figure of a lady attired in tunic and mantle, with the wimple around her
face, holding up in her right hand a shield bearing an orle, Balliol, and in her left
one charged with a lion rampant, crowned, Galloway.   Beneath these, on each
side, suspended from a tree, is a shield, that on the dexter bearing three garbs,
Chester (derived from her grandmother, Maud, daughter of Hugh Earl of
Chester).   The shield, on sinister side, bears two piles in point, Huntingdon
(derived from her grandfather, David Earl of Huntingdon).   *Three piles are*
the proper charge of Huntingdon ; the omission here of one is probably an error
of the seal engraver.
S' DERVORGILLE DE BALLIOL FIL. ALANI DE GALEWAD."

72. COUNTER SEAL of the Last. *Plate V. fig. 2.*
This, like the preceding, is pointed oval, and has three shields suspended from a tree ;
the centre and largest one bears Galloway as in the last, dimidiating Balliol.
The two others bear Chester and Huntingdon as before.
" S' DERVORGILLE DE GALEWAD DNE DE BALLIOLO."—*Appended to the Foundation*
Charter of Balliol College, Oxford, A.D. 1282.

73. BALLIOL, ALEXANDER

A fine seal, in excellent preserva
tion. An armed knight on horseback. The shield on his left arm
bearing an orle, which is repeated
on the housings.

"S' ALEXANDRI DE BALLIOLO."—*Appended to General Release given
by John Balliol to Edward I., 2d
January 1292.—H.M. Record
Office.*

74. BALLIOL, ALEXANDER. THE SAME
PERSON AS THE LAST.

An orle. A wyvern or lizard at each
side of the shield, and foliage at
the top.

"S' ALEXANDRI DE BALLIOLO."—*Ap-
pended to Homage Deed, 10th July 1295. H.M. Record Office.*

75. BALLIOL, WILLIAM.

An orle.

"S' WILLMI DE BALIOL." *Detached Seal. H.M. Record Office.*

76. BALLIOL, WILLIAM.

A stag's head caboshed; between the antlers a shield bearing an orle. A small
crescent and a star at the sides.

"S' WILLMI DE BAYLOL." *Detached Seal.—H.M. Record Office.*

77. BALNAVES, HENRY.

Per fess, a chevron counterchanged. In base a cinquefoil; around the shield the
initials M-H-B.

"S' MAGISTRI HENRICI BALNAVIS."—*Appended to Treaty for Marriage between
Edward VI. and Mary Queen of Scots, 1st July 1543.—H.M. Record Office.*

78. BALNERATH, HUGH.

A very curious device; a fox on its hind legs, holding a shepherd's crook in its fore
paws, in front of a tree in which a bird is perched.

" 81. HUGO DE BALNERATH."—*Appended to the same Homage as No. 48, A.D. 1292.*
— H.M. Record Office.

79. BANE, JOHN.

Rudely executed. A dog's (?) head couped, contourné.

" S' JOHNIS BNOE."—*Appended to Resignation of the five pound lands of Long
Newton into the hands of Sir Alan Stuart, Knight, Lord of Ochiltree, A.D. 1392.
— Morton Charters.*

80. BANNATYNE, JAMES, BURGESS OF EDINBURGH, ANCESTOR
OF GEORGE BANNATYNE, IN HONOUR OF WHOM THE BAN-
NATYNE CLUB WAS NAMED.

A cross fimbrated (?), cantoned with three mullets and a
crescent, the latter in the sinister canton.

" SIGILLUM JACOBI BANNATYNE."—*Appended to a Precept
of Clare Constat to Elizabeth Charteris, of the lands of
Kincleven and others in Perthshire, 6th August 1549. — Breadalbane Charters.*

81. BANNERMAN, JOHN, OF ALSICK.

Three bears' heads erased.

" S' JONIS BANERMAN DE ALSIC."—*Appended to Charter of the lands of Cruives
(" Smythalch, moderné dict. Cruives"), A.D. 1446.—City of Aberdeen Charters.*

82. BARCLAY, WALTER, OF TOWIE.

A chevron between three crosses patté.

" S' WALTERI BARCLA."—*Appended to Precept of Seisin by John Cheyne of
Straloch, for infefting Alexander Cheyne, his nephew, and Katherine Meldrum,
spouse of Alexander Cheyne, in certain lands, 4th July 1499.—Communicated
by Lord Lindsay.*

83. BARCLAY, DAVID.

On a chevron between two crosses patté in chief, and a crescent in base, a mullet.

" S' DAVID BARCLAY," A.D. 1506.— *Marischal College, Aberdeen.*

84. BARTHOLOMEW, WILLIAM, BURGESS OF EDINBURGH.

Ermine, a cross engrailed, surmounted in the centre with an escallop shell.

" S' WILL[ELMI BARTHO]LUMEU."—*Appended to Charter by W. Bartholomew of the
land called Quarrelpotts, at the Greyfriars, Edinburgh, to Roger de Boswell,
circa 14th century.—Edinburgh Charters.*

85. BARTRAM, ANDREW, Prebend of the College Kirk of St. Mary's in the Fields.

Two bullets paleways on the dexter, and as many ermine spots on the sinister.

" s' ANDRIE BARTREME."—*Appended to Charter by A. Bertram of his tenement of land lying at the Churchyard of the College Kirk of St. Mary's in the Fields, to John Fenton, 20th April 1575. — Edinburgh Charters.*

86. BELCHIS, RICHARD.

A crescent and star, not on a shield.

" s RICARDI DE BELCHIS." – *Appended to the same Homage as No. 48, A.D. 1292 — H.M. Record Office.*

87. BELL, ROBERT.

A shield bearing a chevron between three mullets.

" s ROBERTI BELL."—*Appended to an Inquisition, A.D. 1427. Dean and Chapter of Durham.*

88. BELL, ROBERT.

On a shield, three bells.

" s' ROBERTI BELE." *Appended to Perambulation of the lands of Adam Forman, A.D. 1430. Dean and Chapter of Durham.*

89. BELL, ADAM.

A chevron between three birds (parrots?).

" s' ADAM BELLE." — *Appended to an Agreement between the Prior of Coldingham and some free tenants claiming right of common, 16th April 1434. Dean and Chapter of Durham.*

90. BELL, ADAM.

Three bells. Foliage at the top and sides of the shield.

" s' ADAMI BELL." *Appended to a Charter by Alexander Lauder and Mariote Hay, his spouse, of some property in Edinburgh, to W. Paterson, 3d July 1555. — Edinburgh Charters.*

91. BELL, ADAM.

A star or wheel ornament, not on a shield.

" s ADE BEL." *Detached Seal.—H.M. Record Office.*

92. BELL, ALAN.

A duck passant, not on a shield.
"S' ALANI BEL."   Detached Seal.—*H.M. Record Office.*

93. BELLINGHAM, WILLIAM.

* Oval shape ; a key, not on a shield ; three dots or points at each side.
"S' WILL. DE BELLINGHAM."—*Appended to an Indenture,* A.D. 1285.—*Dean and Chapter of Durham.*

94. BERTON, ROBERT, OF OVERBERNTON.

Per pale, dexter three bars wavy ; sinister a merchant's mark.   Slight foliage at the top and sides of the shield.
"S' ROBERTI BERTON DE OVERBERTOUN."—*Appended to a Charter by Robert Berton of Overbernton to Antony Brissat, of some land or tenements on the south of the High Street of Edinburgh, 4th July 1519.—Edinburgh Charters.*

95. BETON, ANDREW.

A shield charged with a fess.
"S' ANDRE DE BETONE."—*Appended to the same Homage as No. 48,* A.D. 1292.—*H.M. Record Office.*

96. BIKYRTOUN, ROBERT, OF LUFNOIS.

An eagle displayed, the head to sinister.
"S' ROBERTI BIKERTON."—*Appended to Charter by Robert Bikyrtoun of Lufnois to George Hepburne, of the lands of Rollandistoun, 19th September 1456. Ballencreiff Charters.*

97. BINNING OR BYNNYNE, SIMON.   *Plate VII. fig. 1.*

A bend engrailed between a mullet of six points in chief and a crescent in base.   The shield surrounded by tracery.
"S' SIMONIS DE BENY."—*Appended to a Charter by John Etal and Alicia Pynebist, his spouse, to the Convent of the Carmelites, of a payment from a tenement in Aberdeen, 4th July 1399.—Marischal College, Aberdeen.*

98. BISSET, BALDRED.

A fleur-de-lis between two roses or cinquefoils, beneath an arched crown, not on a shield, the seal itself being of that shape.
"SIGILL. BALDREDI DE BISSET."—*Appended to an Acquittance to a Bursar of Durham for £10 Arrears of Pension,* A.D. 1288.—*Dean and Chapter of Durham.*

99. BISSET, WALTER, of Clerkington, Son and Heir of Sir John Bisset.

> On a bend engrailed three escallop shells; in sinister chief point a roundle.
>
> "s. WALTARE BESET."—Appended to Resignation of some Lands in Dumfriesshire, by Walter Bisset to George Earl of March, 14th April 1374.—Buccleuch Charters.

100. BISSET, WILLIAM.

> A peacock passant, not on a shield.
>
> "s' WILELMI BYSHET." Detached Seal.—H.M. Record Office.

101. BISSET, WALTER.

> An eagle with wings expanded, not on a shield.
>
> "s' WALTERI BYSHET." Detached Seal.—H.M. Record Office.

102. BISSET, WILLIAM.

> A boar's head couped, not on a shield. In the background, two mullets.
>
> "s' WILLI BUSET." Detached Seal.—H.M. Record Office.

103. BLAIR, JAMES, Provost of Dundee.

> A chevron between three roundles.
>
> "SIGILL. JACOBI DE BLAIR."—Appended to Quitclaim of an annual rent to Alexander Sinton, A.D. 1462.—Hatton's "Sigilla," p. 118.

104. BLAIR, ALEXANDER, of Balthyock.

> A chevron between three roundles.
>
> "s' ALEX BLAR."—Appended to Sasine of a rig of land in the barony of Rait, in Perthshire, A.D. 1491.—H.M. General Register House.

105. BLAIR, WILLIAM, of Balgillo.

> A chevron between three annulets. In chief a mullet.
>
> "s' GVILIELMI BLAIR DE BALGILLOS."—Appended to Reversion by W. Blair of Balgillo, of the lands of Reidcastle or Inverkeilor, in Forfarshire, A.D. 1607.—H.M. General Register House.

106. BLACKWOOD, WILLIAM, Vicar of Duddingston.

A fess between a mascle in base, a mullet in the dexter, and a crescent in the sinister chief.

" s' dni vilelmi blakwood."—*Appended to a Charter by James Blackwood, brother-german to William Blackwood, Vicar of Duddingston, of the lands of Easter Fedallys, 30th October 1584.—Communicated by W. Fraser, Esq*

107. BLAKLAWES, CATHERINE, Wife of Robert Strachan.

Much broken; three castles (?).

" s' catrina blaklawis."—*Appended to a Reversion. A.D. 1573. Communicated by W. Fraser, Esq.*

108. BOYES or BOIS, THOMAS.

A saltire and chief.

" s' thome de bois."—*Appended to a Fragment of a Homage Deed. A.D. 1292.—H.M. Record Office.*

109. BOIS, WALTER.

A fleur-de-lis of a slender and elegant form, not on a shield.

" sigill. walteri del bois."—*Appended to a Charter by Walter Bois, of three acres of land at Karrederes to St. Cuthbert's of Durham.—Dean and Chapter of Durham.*

110. BOLLESDEN, AGNES.

Oval shape; a star and crescent, not on a shield.

" sigill. agnetis."—*Appended to a Charter by Agnes, "filia Margrete de Bollesden," with consent of Hugo her spouse, of the lands of Bollesden to St. Cuthbert's at Durham.—Dean and Chapter of Durham.*

111. BOLLESDEN, THOMAS.

Oval shape; device of an archer shooting.

" sigill. tome de bolesdune."—*Appended to Confirmation of lands at Bollesden to St. Cuthbert's at Durham.—Dean and Chapter of Durham.*

112. BOLLESDEN, THOMAS.

Oval shape; on a shield, six mascles—three, two, and one.

" sigillum tomas de bollesdune."—*Appended to an Agreement regarding Fishings on the Tweed, A.D. 1250.—Dean and Chapter of Durham.*

113 BOLLESDEN, RALPH

A fleur-de-lis, not on a shield.

" SIGILLUM [RADULPHI] BOLLESDENE."—*Appended to the same Document as the last.*

114. BONAR, WILLIAM.

A saltire; in base a crescent reversed.

" S' WILL. BONAR."—A.D. 1461. *Dun Charters.*

115. BONAR, WILLIAM, OF ROSSY

A saltire couped; in base a crescent.

" S' VILELMI BONER."—*Appended to Procuratory of Resignation of the half of Kincleria, etc. " maid be Elizabeth Charteris, with consent of W. Boner her spouse, in favour of Alison Charteris her sister."—6th August 1549.—Breadalbane Charters.*

116. BONCLE, JOHN.

On a chief, three buckles; the shield supported by an angel with wings expanded, and surrounded by clouds; a pretty seal, but much broken.

" JOHANNIS BONCLE."—*Appended to Indenture between King Henry VI and King James II, to abstain from War, 3d July 1449.—H.M. Record Office.*

117. BONKIL, MARGARET.

A chevron between three buckles, tongues erect.

" S' MARGARETE BONKIL."—*Affixed on Paper dated A.D. 1557.—In the Lyon Office.*

118. BONKIL, JOHN.

On a bend, three buckles, tongues erect.

" S' JOHANNIS BONKYLL."—*Communicated by Albert Way, Esq.*

119. BONKIL, RALPH.

A knight on horseback galloping to the sinister; a sword in his right hand, and on his left a kite-shaped shield, quite plain.

" SIGILL. RANULFI DE BONEKIL."—*Appended to a Charter by R. Bonkil, of land at Edenham, to the Prior of St. Ebbe at Coldingham.—Dean and Chapter of Durham.*

120. BORTHWICK, ALEXANDER, Burgess of Edinburgh.
> The initials A·B· between three cinquefoils.
> " s' ALEXANDRI BORTHYIC."—*Appended to Sasine of lands at Ballencrieff to Robert Borthwick, 9th May 1511.—Blackbarony Charters.*

121. BORTHWICK, WILLIAM, fourth Lord.
> Three cinquefoils.
> " s' WILLELMI DNI BORTHYK."—*Appended to Answer of Refusal to the King of England's advice to remove the Duke of Albany from the government of the young King (James V.) of Scotland, 3d July 1516 — H.M. Record Office.*

122. BORTHWICK, WILLIAM, Lord. Same Person as the last.
> Three cinquefoils; crest on helmet a horse's head; supporters, two lions sejant.
> " s' WILLMI DE BORTHWICK."—*Appended to Precept of Sasine by Lord Borthwick in favour of George Lord Home, of the lands of Hostrotts, Roxburghshire, 26th September 1522.—Home Charters.*

123. BOSWELL, EUSTACE.
> A saltire, and on a chief a boar's head couped.
> " s' EUSTACH DE BOISWEL."   Detached Seal.—H.M. Record Office.

124. BOTEVILLE, RICHARD.
> A lion passant, not on a shield.
> " SIGILL. RICARDI DE BOTEVILLE."—*Appended to a Document relating to Fishings on the Tweed, A.D. 1250.—Dean and Chapter of Durham.*

125. BOTILE, ALEXANDER.
> A fleur-de-lis, not on a shield.
> " s' ALEXANDR BOTILE."   Detached Seal.—H.M. Record Office.

> BOTHWELL, Earl of.   *Vide* Hepburn.

126. BRANXTON, CHRISTIAN.
> A fleur-de-lis, not on a shield.
> " CHRISTIANE FILIE THEOBALDI DE BRANXTON."—*Appended to an Instrument in the Archives of the Dean and Chapter of Durham.*

127 BRANXTON, CHRISTIAN.

    Oval shape : a flower, not on a shield.

       " s' CHRISTIANE DE BRANKSTU."—*Appended to Confirmation by Christian Branxton,*
        " QUONDAM UXOR ROBERTI MARSHALL." *of lands at Bollesden, to the Monks of*
        St. Cuthbert's of Durham.—Dean and Chapter of Durham.

128 BRANXTON, RALPH.

    Two pairs of horse barnacles, not on a shield.

       " SIGILLUM RAUL FILII GILBERTI."—*Appended to Confirmation of the Church of*
        Egwell to St. Cuthbert's, Durham.—Dean and Chapter of Durham.

129 BRANXTON, RALPH.

    A single pair of horse barnacles, not on a shield.

       " SIGILLUM RAUL FILII GILBERTI."—*Appended to a Charter of a carucate of land at*
        Bollesden, to St. Cuthbert's of Durham.—Dean and Chapter of Durham.

130. BRANKSTON, WILLIAM.

    A pair of horse barnacles between a crescent and an estoile, not on a shield.

       " SIGILL. WILELMI DE BRANKIST."—*Appended to Confirmation of Charter by Alex-*
        ander de Bollesden, of some lands at Bollesden, to St. Cuthbert's of Durham.
        —Dean and Chapter of Durham.

131. BRAYVILLE, ROBERT.

    A shield bearing a bull's head cabossed.

       " s' ROBTI DE BRAYVILLA." Detached Seal.—*H.M. Record Office.*

132. BRECHIN, JOHN.

    Oval shape ; a curious device, not on a shield. A fish, and below it a rude-shaped
    hook.

       " s' JOHIS BRICUN."—*Appended to an Indenture,* A.D. 1285.—*Dean and Chapter*
        of Durham.

133. BRECHIN, DAVID.

    A shield bearing three pales.

       " s' DAVIT DE BRECHIN."—*Appended to Homage Deed,* A.D. 1292.—*H.M. Record*
        Office.

134. BRITTANY, MARGARET, Duchess
of, Sister of William the
Lion, and Wife of Conan, Duke
of Brittany. Died, A.D. 1171.
Oval shape; a front figure of a lady
in flowing robes, her right hand
extended, holding a small globe
crowned with a fleur-de-lis, in
her left a falcon.

"sigillum marga[rete britta]norum
ducisse."—*Appended to a Grant
of Land in Forset, Yorkshire,
by the Duchess of Brittany, to
Engeram, her cupbearer.—Vide
"Archæologia Æliana," octavo
series, vol. ii. p. 10.—Sir John
Lawson, Bart., of Brough Hall,
Yorkshire.*

135. BRODIE, DAVID.
A human heart, and in chief two mullets.
"s' dave brode."—*Appended to Charter by David Brodie, burgess of Nairn, to
David Gibson, burgess of said burgh, of a rood of land in Nairn, 22d January
1511.—Cawdor Charters.*

136. BROUN, PATRICK, of Colstoun.
A lion rampant. Foliage at the top and sides of the shield.
"s' patrici broun de colstoun."—*Appended to a Charter by Patrick Broun
of Colstoun, to William Broun, natural son of the deceased William Broun,
alias Coupar, and Elizabeth Darg, his spouse, of certain lands within the Burgh
of Haddington, 25th October 1574.—Colstoun Charters.*

137. BROUN, PATRICK, of Colston.
A lion rampant. Crest, a hand holding a dagger.
"s' patrick broun of colsto."—*Appended to Charter by Patrick Broun to Alan
Martin, of certain lands, 21st January 1597.—Colston Charters.*

138. BROWN, ROBERT, Bailie of Inverkeith.

On a chevron, between three fleur-de-lis, an annulet.

"S' ROBERTUS BROUN." A.D. 1588.—*W. B. D. D. Turnbull, Esq.*

139. BRUCE, ROBERT.

A fleur-de-lis, on which two birds are sitting; not on a shield.

"SIGILLUM ROBERTI DE BRUS."—*Appended to a Grant of the Liberties of Hartlepool to St. Cuthbert's of Durham, circa* 1158.—*Dean and Chapter of Durham.*

140. BRUCE, ROBERT, Son of the above Robert Bruce. *Plate IV. fig. 3.*

A fleur-de-lis, not on a shield.

"SIGILLUM ROBERTI JUVENIS DE BRUS."—*Appended to a Grant of the Chapel of Eden, to the Abbot and Monks of St. Cuthbert's of Durham.—Dean and Chapter of Durham.*

141. BRUCE, PETER, Lord of Skelton.

A knight on horseback, armed in chain mail, barrel helmet, a sword in his right hand, and on his left a shield bearing a lion rampant.

"SIGILLUM PETRI DE BRUS."—*Appended to Confirmation by the Lord of Skelton, of the Charter of W. and R. Brus, of the Liberties of Hartlepool to St. Cuthbert's of Durham.—Dean and Chapter of Durham.*

142. BRUCE, MARGARET, Lady of Kendal, Third Daughter and Co-heiress of the above Peter de Bruce, Lord of Skelton.

A pretty oval-shaped seal. A full-length figure of a female in an ermine-lined robe, supporting in her right hand a shield, charged with three water bougets the coat of Ross, and in her left her paternal coat, the lion rampant of Bruce.

"SIGILLUM MARGARETE DE ROS."—*Appended to an Indenture between Lady Margaret de Ross and William de Stirkland, A.D.* 1280.—*From the Collection of the late W. B. D. D. Turnbull, Esq.*

D

143. BRUCE, ROBERT, Earl of Carrick.

  A fine example of the art of the period. A lion passant
  gardant, not on a shield; the background ornamented
  with foliage.

  "SECRETUM SECRETORUM."—*Appended to the Homage Deed
  of Robert Bruce, William Douglas, and Alexander Stuart.
  A.D. 1296.—H.M. Record Office.*

144. BRUCE, ROBERT.

  A fragment of a fine seal. A knight on horseback; on his left arm a shield bearing
  a saltire and chief.—*From the Collection of Casts taken by the late Mr.
  Doubleday, by whom it is marked A.D. "1297," but it is perhaps earlier.*

145. BRUCE, ISABELLA, Daughter of Robert Bruce, Earl of
  Carrick, Sister of King Robert the Bruce. Married
  Thomas Randolph, by whom she had the celebrated
  Thomas Randolph, Earl of Moray.

  An elegantly designed seal; pointed oval shape: a full-length
  front figure of a lady, her right hand extended, holding
  a fleur-de-lis apparently (the impression is rather indistinct), her left hand on her breast; the background ornamented with foliage.

  "SIGILLU. ISABELLE DE BRUS."—*Communicated by Albert
  Way, Esq.*

146. BRUCE, JOHN.

  Oval shape; a hawk preying on a bird, amid foliage, not on a shield.
  "SIGILLUM JOHIS BRUS." Detached Seal.—*H.M. Record Office.*

147. BRUCE, ROBERT.

  A saltire, and on a chief three mullets.
  "S' ROBERTI BRUS," A.D. 1495.—*Communicated by D. Laing, Esq.*

  BUCCLEUCH, Earl of. *Vide* Scott.

148. BUCHAN, THOMAS.

  A flower or wheel ornament, not on a shield.
  "S' THOMAE DE BUCAN." Detached Seal.—*H.M. Record Office.*

149. BULLYN, THOMAS, Canon of Glasgow.

A bull's head caboosed. The shield supported by an angel with wings expanded
" SIGILLUM DNI THOME BULLYN," A.D. 1460.—*Communicated by W. Fraser, Esq.*

150. BURDEN, WALTER.

A five-leaved flower or wheel ornament, not on a shield
" S' VALTERI DE BURDEN."—*Appended to the same Homage as No. 48, A.D. 1292.—
H.M. Record Office.*

151. BURDUN, ROGER.

Oval shape; a device of a bird, with ears of grain.
" S' ROGERI DE BURDUN." Detached Seal.—*H.M. Record Office.*

152. BURNAL, WILLIAM.

A flower or wheel ornament, not on a shield.
" S' WILL. BURNAL." Detached Seal.—*H.M. Record Office.*

153. BURNVILE, ARCHIBALD.

A small oval seal, with the Virgin sitting with the infant Jesus in her arms; in the
background a lamp.
" S' ARCHE BALDI BURNVILE." Detached Seal. *H.M. General Register House.*

154. BUTLER, JOHN

Oval shape; a rose or wheel ornament, not on a shield.
" S' JOHANNIS DE BUTELER." Detached Seal.—*H.M. Record Office.*

155. CAIRNS, WILLIAM, Vicar of Glammis.

A mullet of six points; on a chief, three birds. Above the shield a demi figure of
the Virgin and Child. Supporters, two lions sejant gardant.
" S' WILLELMI DE CAIRNS."—*Appended to an Indenture between Walter and
Thomas Ogilvie and others, 25th September 1455.—Sir John Ogilvy, Bart.*

156. CALDECOTE, GEOFFRY.

A head in centre of tracery.
" S' GALFRIDI DE CALDECOTE." Detached Seal.—*H.M. Record Office.*

157. CALDECOTE, JOHN.

A saltire, and on a chief three escallop shells.
" S' JOHANNIS CALDECOT."—*Appended to Charter by John of "Caldecotys" to William*

of " Caldecotys," his Son, and Christian, Daughter of Walter Tweedy, his Spouse, of the lands of Graydon in the Earldom of March, and the lands of Symprync in the shire of Berwick. Not dated, but probably about A.D. 1360.—H.M. General Register House.

158.—CALDECOTE, CHRISTIAN, WIDOW OF WILLIAM CALDECOTE.

A saltire, and on a chief a mullet and two escallop shells, the mullet in the dexter.

" S' CRISTIANE DE CALDECOTTIS." The first name has been evidently Katherine, but altered to Christiane.—Appended to Renunciation of the lands of Grayden and Symprync by Christian of Caldecote, A.D. 1424.—H.M. General Register House.

159. CALDECOTE, ELIZABETH, WIFE OF WILLIAM LIVINGSTON OF BALCASTEL, for whose seal vide No. 535 of " Descriptive Catalogue of Scottish Seals."

A saltire, and on a chief three escallop shells.

" S' ELIZABETH DE LIVINGSTON."—Appended to Resignation of Symprync, etc., 8th November 1459.—H.M. General Register House.

160. CAMBRUN, ROBERT, OF BALEGRENACH.

Three bars, and a label of five points.

" S' ROB. CAMBRUN D. BALNEGU." Detached Seal.—H.M. Record Office.

161. CAMPBELL, ARCHIBALD, SECOND EARL OF ARGYLE.

Quarterly, first and fourth, gyronny of eight for Campbell; second and third, a galley (lymphad) for Lorn. Crest on helmet, a boar's head and neck couped. Supporters, two lions.

" S' ARCHI COM. ERGADIE DNI [CAMPBELL] ET LORN."—Appended to Precept of Sasine for infefting Ewan Mackorcadel in lands in Argyle, in exchange for the lands of Edderlyn, 18th July 1495.—Argyle Charters.

162. CAMPBELL, SIR JOHN, KNIGHT, OF CAWDOR, THIRD SON OF THE SECOND EARL OF ARGYLE, ANCESTOR OF THE EARL OF CAWDOR, BY MARRIAGE WITH MURIEL, SOLE HEIRESS OF THE ANCIENT THANES OF CAWDOR.

Couché; gyronny of eight, with a mullet as a difference; helmet and mantling; the crest is lost.

" S' DNI JOHANIS CAMBAL DE [CALDER]MIL." Detached Seal.—Cawdor Castle.

163. CAMPBELL, SIR JOHN, KNIGHT, OF CAWDOR. SAME PERSON AS THE LAST.

Much damaged, but evidently quarterly: first and fourth Campbell, second Lorn, third Cawdor.

" s' JOHANNIS CAMPBEL DE CALDOR MILITIS."—*Appended to Procuratory by Muriel de Cawdor and Sir John Campbell, her husband, for resigning all their lands in favour of themselves, 17th July 1511.—Cawdor Charters.*

164. CAMPBELL, COLIN, THIRD EARL OF ARGYLE, SON OF THE ABOVE ARCHIBALD EARL OF ARGYLE.

The same design and charges as No. 161, but in this the boar's head is erased, and the background foliated.

" s' COLINIS COMES ARGADIE."—*Appended to Precept of Clare Constat in favour of Ewan Mackorcadel, 24th March 1518.—Argyle Charters.*

165. CAMPBELL, ARCHIBALD, AFTERWARDS FIFTH EARL OF ARGYLE.

Quarterly, Campbell and Lorn.

" s' ARCHIBALDI MAGISTRI ARGA."—*Appended to a Charter by the Earl and Master of Argyle of the lands of Kanderagan to Duncan Macgillespalk, and Patrick M'Kellar, his natural son and heir-male, 8th December 1545.—Rev. Dr. M'Kellar.*

166. CAMPBELL, ARCHIBALD, THE SAME PERSON AS THE LAST.

Quarterly, Campbell and Lorn. Crest on helmet, a boar's head couped : supporters, two lions. Very rude workmanship.

" s' ARCHIBALDI M. DE ARGADIE."—*Appended to Precept of Sasine of the lands of Kilbryde in favour of Colin Campbell, 14th September 1550.—Breadalbane Charters.*

167. CAMPBELL, ARCHIBALD, SAME PERSON AS THE LAST.

Quarterly, Lorn and Campbell.

" s' ARCHIBALDI CABEL M. DE ARGYL."—*Appended to an Indenture between the Earl and Master of Argyle and Sir Colin Campbell of Glenurquhay, 25th March 1553.—Breadalbane Charters.*

168. CAMPBELL, ARCHIBALD, EIGHTH EARL OF ARGYLE. Executed at Edinburgh, May 1661.

Quarterly, Campbell and Lorn. Crest on front helmet, a boar's head erased ; supporters, two lions rampant. Foliage fills the background.

" s' ARCHIBALDI COMITIS ARGADII ET DO. LORN."—*Appended to a Charter among the Family Charters of Campbell of Stonefield, Argyleshire.—Communicated by E. Colquhoun, Esq.*

169. CAMPBELL, COLIN, of ARDKINGLASS.
    Gyronny of eight.
    " S' COLENE CAMBAL."—*Appended to the same Instrument
    as No. 165.*

170. CAMPBELL, COLIN, THE SAME PERSON AS THE LAST.
    Plate III. *fig.* 4.
    A very neat small seal. Gyronny of eight, with a mul-
    let in centre as a difference.
    " S' COLYNE CAMPBELL OF ARDKYNGLAS."—*Appended to the same Instrument as
    No. 166.*

171. CAMPBELL, SIR COLIN, KNIGHT, OF GLENURQHAY, ANCESTOR OF THE MARQUIS
    OF BREADALBANE. *Plate* VIII. *fig.* 4.
    Quarterly, first and fourth, Campbell; second, Lorn; third, a fess chequé for Stuart.
    Crest on helmet, a boar's head and neck couped; supporters, two antelopes.
    Background ornamented with foliage.
    " S' COLINI CAMPBEL DE GLENURQUHAY."—*Appended to Precept of Sasine by Sir
    Colin Campbell, of six marks from the lands of Auchinryre, etc., in favour of
    Patrick Campbell, his son. A.D. 1556.—Breadalbane Charters.*

172. CAMPBELL, DUNCAN, FIAR OF GLENURQUHAY. SON OF THE ABOVE SIR COLIN.
    Plate III. *fig.* 5.
    Quarterly, as in the last. Foliage at the top and sides of the shield.
    " S' DUNCANI CAMPBELL."—*Appended to a tack of Auchnacraig and Killin by the
    Chancellor (Dougall Macgregor) and Chapter of Lismore, to Patrick Camp-
    bell, 28th December 1574.—Breadalbane Charters.*

173. CAMPBELL, NIGEL, RECTOR OF KILMARTIN AND PRECENTOR OF LISMORE, AFTER-
    WARDS BISHOP OF ARGYLL.
    Gyronny of eight; a mullet in centre for difference.
    " NIGELLI CAMPBEL "—*Appended to the same Instrument as the last.*

174. CAMPBELL, WILLIAM.
    Oval shape. The Virgin and child sitting; at each side a rose.
    " S' WILLELMI DE CAMBELL CLERICI."   Detached Seal.—*H.M. Record Office.*

175. CAMPBELL, JAMES. of Thornton.

Quarterly, Campbell and Lorn. Crest on helmet, with mantlings, a boar's head and neck.

" S' JOHANNIS CAMPBEL DNI DE THORNTON." Detached Seal.—*H.M. Record Office.*

176. CANT, ALEXANDER.

A bend engrailed between two crescents.

" S' ALEXANDER CANT." Detached Seal.— *Museum of Society of Antiquaries of Scotland.*

177. CARBAR, JOHN.

A fleur-de-lis, not on a shield.

" SIGILL. JOHANNIS DE CARBAR." A.D. 1290.—*Hutton's " Sigilla,"* p. 186.

178. CARLTON, DUNCAN.

A squirrel sitting, not on a shield.

" S' DUUCANI DE CANLT." Detached Seal.—*H.M. Record Office.*

179. CARNCROSS, or CAIRNCROSS, WILLIAM. of Colmbray.

A stag's head couped.

" S' WILELMI CARNCROS "—*Appended to a Sasine, dated 23d June 1545.—Edinburgh Charters.*

180. CARNEGIE, JOHN. of Kinnaird.

An eagle displayed standing on a barrel, with a mullet in the dexter and sinister chief points.

" S' JOHANNIS DE CARNAGE."—*Appended to Sasine of the lands of Brecky and Ballysham to Lord Ogilvy, by David, Abbot of Arbroath.* A.D. 1489.—*Greenhill Charters.*

181. CARNEGIE, Sir ROBERT, Knight of Kinnaird. Grandson of the last.

An eagle displayed surmounting a barrel. Foliage at the top and sides of the shield.

" S' [ROBERTI] CARNEGY."—*Appended to Precept of Clare Constat to James Deanistoun, of the lands of Loch-mgle, by Sir Robert Carnegie, Baron of the Barony of Carrisden.* A.D. 1551. *Communicated by J. T. Gibson-Craig, Esq.*

The family of Kinnaird are said to have been from an early period Cupbearers to the King, and as indicative of the office carry a covered cup as one of their charges, placing it on the breast of the eagle, as in the following seals. The two preceding examples, however, prove that previous to the middle of the sixteenth century the cup had no place on the family coat, and the barrel having a more appropriate allusion to the office of Cellarer or " Butelarius" than of Cupbearer, it seems probable that in early times the former office was held by the family. This view appears to receive some confirmation from a Charter of James VI., dated at Holyrood, 14th October 1491, in the archives at Kinnaird Castle. This Charter is given to Sir John Carnegie in liferent, and to his brother, David Carnegie, of Culluthye, who carried on the line of the family, erecting the lands of Kinnaird, with manor, fortalice, etc. etc., into the Barony of Kinnaird, to be held by the tenure of a penny Scots yearly, if demanded, and keeping the King's cellar of ale whenever the Court resided in the country.

182. CARNEGIE, SIR JOHN, KNIGHT, SON OF THE ABOVE SIR ROBERT. *Plate* VIII. *fig.* 7.

An eagle displayed, charged with a covered cup; foliage at the top and sides of the shield.

" S᾽ D. JOANNIS CARNEGY DE KYNARD MI." The whole is but rudely executed, but the cup appears here for the first time in place of the barrel, whether in allusion to the office of " Butelarius " or Cupbearer.—*Appended to Resignation by Sir John Carnegie of lands in Aberdeen, etc., in favour of David Carnegie of Culluthye, his brother, 16th September 1591.—Carnegie Charters.*

183. CARNEGIE, DAVID, CREATED EARL OF SOUTHESK 22D JUNE 1633. *Plate* VIII. *fig.* 8.

An eagle displayed, charged with a covered cup; above the shield an earl's coronet and helmet, with mantling; crest a thunderbolt; motto on ribbon, " DREAD GOD ;" supporters, two greyhounds.

" SIGILLUM DAVIDIS COM. DE SOUTHESK DOM. CARNEGY DE KYN. ET LEUCHARS." — *Appended to Charter by the Earl, of lands in Forfar, to Andrew Lyel, 10th April 1655.—Carnegie Charters.*

184. CARRICK, DUNCAN, EARL OF.

A dragon passant, not on a shield.

" SIGILL[UM DUNCANI COMITI]S DE CARRIC."—*Appended to Charter of the patronage*

of the church of Maybole to the Nuns of North Berwick by Duncan Earl of
Carrick.—From the Collection of General Hutton, who gives the above reference,
and adds, "Original Charter at Panmure."—Hutton's "Sigilla," vol. ii. p. 93.

185. CASSIE, ANDREW.

A saltire between two crescents in the flanks, and as many mullets in base and
chief.

"s' AND. CASSIE." A.D. 1594.— *The late P. Chalmers, Esq.*

186. CATHCART, JOHN, of Glendowys.

A fess between three cross-crosslets fitché; in honour point, a human heart

"s' JOHANNIS CATHKERT."—*Appended to Precept of Sasine for infefting Hugh Earl*
of Eglinton in the lands of Drumlonford, in Ayrshire, 24th February 1511.—
Eglinton Charters.

187. CATHCART, SYBILLA, Wife of the above John Cathcart, and daughter
" Dni Alani Cathcart " of Carleton.

A stag's head contourné, between three cross-crosslets fitché.

"s' CYBEL CATHKART."—*Appended to the same Instrument as the last.*

188. CATHCART, ALAN.

Oval shape. The Virgin and Child.

"SECRETU ALANI CATH . . . ."—*Appended to a fragment of a Deed of the thirteenth*
century, probably a Homage Deed. H.M. Record Office.

189. CAWDOR, WILLIAM, Thane of, Shiriff of Nairn.

Very imperfect. A stag's head cabossed, between the horns a buckle ?,. Apparently
a chief with a mullet on the sinister point. (These are nearly the same charges
as those of his father, Donald Calder. See No. 150 of " *Descriptive Catalogue*
of Scottish Seals.") The legend is lost.—*Appended to Precept of Sasine. A.D.*
1449.—H.M. General Register House.

190. CAWDOR, WILLIAM, same person as the last.

A stag's head cabossed, between the antlers a buckle;
the tongue fesswise; a mullet in the dexter and
sinister chief points; branches of foliage are also
in the field, being doubtless a rude imitation of
the diapering of a previous century.

" SIGILLUM WILLIMI DE CAWDOR."—*Appended to Pro-*
curatory of the lands of Balmakeith, in Nairn.
A.D. 1458.— Cawdor Charters.

E

191. CAWDOR, MURIEL, Granddaughter and sole Heiress of the above William Cawdor.

> The shield, of rather unusual shape, is so much damaged that it cannot be satisfactorily blazoned. It seems to be a fess between two mullets in chief, and a stag's head caboshed in base.

> "s' mureelle caldor dna de eode."—*Appended to the same Instrument as No. 162.*

192. CAWDOR, DAME MURIEL, same Lady as the above.

> A stag's head caboshed, and on a chief an oval buckle, the tongue palewise, between two mullets.

> "s' murielle calder."—*Appended to Procuratory of Resignation of all her lands and baronies in favour of John Campbell, son of Archibald Campbell, "her oye," 9th June 1573.—Cawdor Charters.*

193. CAWDOR, HUGH, Sheriff of Nairn, third son of the above William Cawdor.

> A stag's head caboshed, the antlers within a circle, which may perhaps be meant for a buckle: it is not on a shield, and therefore is merely some fanciful device of the owner.

> "s' hugonis calder."—*Procuratory of Resignation of the Sheriffship of Nairn, and Constabulary of the Castle of Nairn, to Sir John Campbell, 16th December 1528.—Cawdor Charters.*

194. CAWDOR, CAMPBELL of.

> A stag's head caboshed, between the antlers a buckle, tongue fesswise. Rudely executed.

> "s' campbell of kaudor deope." Detached Seal.—*Cawdor Castle, probably seventeenth century.*

195. CATHKENE or CATHKIN, MARGARET, Wife of David Kinloch, one of the Ministers of Edinburgh.

> A saltire, in chief a mullet.

> "s' margareta cathkin."—*Appended to Reversion by David Kinloch, and Margaret Cathkene his spouse, of four acres of land at Lochbank, near Edinburgh, 5th April 1550.—Edinburgh Charters.*

196. CHALMERS, THOMAS, Bailie of Aberdeen.

Conché; a demi-lion issuing from a fess; in base a fleur-de-lis; above the shield a helmet; crest now lost.

"s' thomas de chalmeri."—*Appended to "Sasine of four-shillings' rent to Thomas Chalmers, Curate of Aberdeen."—From the Collection of General Hutton, who gives the above reference, but no date.*

197. CHALMERS, ALEXANDER, of Murthill.

Conché, as in the last; crest on helmet, a bird devouring its prey; supporters, two lions rampant.

"s. alexand. de cam."—*Appended to Charter by Alexander Chalmers, of tenements in the old road south of the highway, Aberdeen, 6th December 1449.—Marischal College, Aberdeen.*

198. CHALMERS, ALEXANDER, one of the Prebends of the College Kirk of St. Mary-in-the-Fields.

A fess chequé between two estoiles in chief, and a crescent in base.

"s' m. alexander chalmair."—*Appended to Charter by Alexander Chalmers to John Fenton, Comptroller's Clerk, of a tenement, etc., 10th April 1575.—Edinburgh Charters.*

199. CHALMERS, ROBERT.

A lion rampant, not on a shield.

"s' robti camlra." Detached Seal.—*H.M. Record Office.*

200. CHALMERS, WILLIAM.

Curious device. Not on a shield, a lion devouring the leg of a horse (?).

"s' wilelmi de camera." Detached Seal.—*H.M. Record Office.*

201. CHAMPAGNE, PEIRS, Parson of the Church of Kynkell.

Fretty.—*Appended to Homage Deed, 4th July 1295.—H.M. Record Office.*

202. CHAPMAN, DAVID.

A saltire engrailed, in chief a boar's head and neck couped.

"s' david chepman." Detached Seal, probably about 1480.—*Coiston Charters.*

203. CHARTERS, WILLIAM.

A six-leaved flower, not on a shield.

"s' will. d' chartris."—*Appended to the same Homage as No. 48. A.D. 1292.— H.M. Record Office.*

204. CHARTERS, ROBERT.

Oval shape, a star or wheel ornament.

"s' roberti d' chartris."   Detached Seal.—*H.M. Record Office.*

205. CHATOU, ADAM.

A fleur-de-lis, not on a shield.

"s' ade de chatou."—*Appended to the same Homage as No. 48. A.D. 1292.—H.M. Record Office.*

206. CHIRNSIDE, MARIOTE.

Oval shape. A tree or bush, not on a shield.

"si' mariete de chirneside."—*Appended to a Charter by Mariote Chirneside, "quondam uxor Ricardi de Reston," of a caracute of land at Remingston to the Prior of St. Ebbe at Coldingham.—Dean and Chapter of Durham.*

207. CHISHOLM, RICHARD.

A boar's head, couped, contourné.

"s' ricardi de chiselm."—*Appended to the same Homage as No. 48, A.D. 1292.—H.M. Record Office.*

208. CHISHOLM, MURIEL, Wife of Alexander Sutherland of Duffus.

Per pale, dexter; three cross-crosslets fitché, for Cheyne, and in chief three mullets for Sutherland of Duffus; sinister, a boar's head erased, and a chief for Chisholm.

This is an interesting seal, as showing the introduction of the coat of Chisholm to the Duffus arms. Here they are properly placed on the sinister side, impaled with the arms of her husband, Sutherland of Duffus, namely, the mullets of Sutherland, and the cross-crosslets of Cheyne. The three coats were subsequently carried tierced, as we find in the seal of William Sutherland, A.D. 1540, afterwards to be described, the fourth in descent from this marriage; at a later period they were quartered, first and fourth Sutherland, second Cheyne, third Chisholm, and thus continued till finally superseded by the coat of Dunbar, on the succession of Sir Benjamin Dunbar of Hempriggs, in 1827, as heir of line to the title.

"s' murielle chisholm."—*Appended to a Writ by M. Chisholm, with consent of Alexander Sutherland of Duffus, her spouse, constituting certain persons named her proxies for resignation of her lands in Paxton, with the fishings in the waters of the same, to St. Ebbe at Coldingham. 19th March 1435.—Dean and Chapter of Durham.*

209. CHISHOLM, WILLIAM.

A boar's head and neck, couped, contourné.

"s' VILAMI CHISHOLME DE KINKELL."—*Appended to Charter by William Chisholm, fiar of Kinkell, to John Mackenzie of Gairloch, A.D. 1592.—Communicated by Lewis Mark Mackenzie, Esq.*

210. CLELAND, ANDREW, ONE OF THE BAILIES OF EDINBURGH.

A hare or rabbit rampant, with a hunting-horn suspended from its neck.

"s' ANDREE KNELAND."—*Appended to a Sasine dated 3d October 1612.—Edinburgh Charters.*

211. COCKBURN, SIR JOHN, KNIGHT.

A cock, not on a shield.

Appended to "Concordia" by the Commissioners for governing the Marches, 12th July 1429.—H.M. Record Office.

212. COCKBURN, ALEXANDER.

Three cocks contourné.

"s' ALEXANDER COKBURN."—*Appended to Charter by Alexander Cockburn to George Brown of Colston, 1519.—Colston Charters.*

213. COCKBURN, PATRICK, OF NEWBIGGING.

Three cocks passant; a crescent in fess point, as a difference; foliage at the top and sides of the shield.

"s' PATRICH COKBURN."—*Appended to Charter by Patrick Cockburn of Newbigging to William Cockburn of Chouslie and Margaret Galbraith his spouse, of the Temple lands in the barony of Langton, co. Berwick, with the corn-mill thereof, 20th June 1539.—David Laing, Esq.*

214. COCKBURN, ROBERT.

Three cocks.

"s' ROBERTI COKBURNE."—*Appended to Truce and Declaration of Thomas Duke of Norfolk, 23d and 29th November 1524.—H.M. Record Office.*

215. COCKBURN, ROBERT.

Much damaged. A bust of the Virgin in a crescent; a shield at each side, the sinister one apparently charged with a cinquefoil.

"s' ROBTI DE COKEBURNE." Detached Seal.—*H.M. Record Office.*

216. COLBAN, MARGARET.

Oval shape. A thunderbolt (?) and three stars, not on a shield.

" s' MARGAT D. COLBAN." Detached Seal.—*H.M. Record Office.*

217. COLDANE, JOHN, CHANCELLOR OF THE CATHEDRAL CHURCH OF BRECHIN.

On a fess, between a trefoil slipt in base, and three ermine spots in chief, two mullets; a scroll ornament at the top and sides of the shield.

" S' M. JOHANNIS COLDEN, CANCALARII."—*Appended to an Ecclesiastical Instrument dated 10th January* 1537.—*Colston Charters.*

218. COLDINGHAM, ADAM.

Oval shape; a very pretty seal. A monk kneeling at prayer before the Virgin and Child.

" s' ADE D' COLDINGHAM."—*Appended to a Charter by Ade, son of Thomas Stuke of Coldingham, of a toft at Westerton to the Priory of Coldingham.—Dean and Chapter of Durham.*

219. COLLACE, ROBERT, OF BALNAMOON.

Three mascles, bendwise sinister, between two saltires couped.

" s' ROBERTI COLLAISE."—*Appended to an Instrument dated A.D.* 1574.—*Findowrie Charters.*

220. COLLACE, JOHN, SON OF THE ABOVE ROBERT.

On a bend, between two saltires couped, three mascles.

" s' JOHANIS COLESE."—*Appended to the same Instrument as the last.*

221. COLQUHON, SIR JOHN, OF LUSS.

A saltire engrailed; crest on helmet, a stag's head; supporters, two greyhounds; behind the sinister one a mullet.

" s' JOHANNIS DE CULQUHON DE CULQN." (?)—*Appended to Receipt by " J. Colquhon, Lord of Luss, to haiv resavit thankful and full payment from the English Ambassadors and Jas. Schaw of Sauchy the sume of* 600 *marks, English money, for a ship captured by Lord Gray,"* 8th May 1475.—*H.M. Record Office.*

222. COLVIL, ROBERT, MASTER OF.

Quarterly: first and fourth, a fess chequé for Stuart; second and third, a cross moline for Colvil.

" s' ROBERTI COLVIL." Detached Seal, about 1576.

223. COLVILE, ROBERT, SECOND LORD COLVILE, OF OCHILTREE.

A saltire, and a chief bearing two mullets; foliage at the top and sides of the shield.

"S' ROBERTI COLVILL DE CLEISH."— *Appended to a Charter by Thomas Bruce of Blairhall to James Sandis, in Strathmilne, and Margaret Braw (?) his spouse, of lands in the regality of Culros, and sheriffdom of Perth. 16th August 1663. In possession of William Johnston Stuart, Esq., the lineal representative of Bruce of Blairhall.*

This seal of Colvile must have been appended by mistake for that of Thomas Bruce, the granter of the charter. The charges of both are very similar, and without attention to the legend such a mistake might easily occur. In the deed it is expressly said, " my proper seal is appended," which, however, it cannot be. The same mistake occurs in another charter by the same Thomas Bruce in the following month (September 1663).

224. COMRIE, JOHN, OF COMRIE.

A chevron between a mullet in base, and two ears of rye (?) chevronwise in chief.

"s' JOANNIS COMRI."—*Appended to Resignation of some office in Strathern.* A.D. 1495. —*Communicated by David Laing, Esq.*

225. COMRIE, JOHN, OF COMRIE.

A bend sinister between two mullets in chief, and one in base.

"s' JHOES COMRIE," A.D. 1530.—*Communicated by David Laing, Esq.*

226. CONAN, DAVID.

A saltire, with a rose in the dexter, and a crescent in the sinister cantons.

"DAVID DE [CONAN]," A.D. 1444.—*Brechin Charters.*

227. CONVETH, JOHN.

Oval shape. The eagle of St. John, with a scroll, on which is inscribed, " s' JOHIS." "s' JOHIS DE CONVETH CLERICI." Detached Seal.—*H.M. Record Office.*

228. COPLAND, THOMAS.

Very rudely executed. Two garbs in chief; a boar's head contourné in base.

"s' THOME COPLAND," A.D. 1524.—*Dun Charters.*

229. COR, CLEMENT, ONE OF THE BAILIES OF EDINBURGH.

A cross engrailed, with a human heart in the first and fourth quarters, and a rose in the second and third.

"s' CLEMENTIS COR."—*Appended to a Sasine in favour of J. Scott, 6th June 1560. —Edinburgh Charters.*

230. CORBETT, ROGER.

An antique gem. A warrior putting on his armour.

"SIGILLUM ROGERI CORBETT."—*Appended to the same Homage as No. 48, A.D. 1292.*
—H.M. Record Office.

231. CORMAN, JOHN.

Device of a bird, not on a shield.

"S' JOHIS DE CORMAN." Detached Seal.—*H.M. Record Office.*

232. CORNHALL, WILLIAM.

A wheat sheaf, not on a shield.

"S' WILL. DE CORNALE."—*Dean and Chapter of Durham.*

233. CORNTON, FLORENCE.

A fess chequé between a rose (?) in chief, and a trefoil slipped in base.

"S' FLORENCE CORNTOUN."—*Appended to a Sasine in favour of Patrick Crombie, and Isabell Moncreiff his spouse, of a tenement at Leith, 1st June 1554.— Edinburgh Charters.*

234. CORNWELL, ELLEN.

Three Cornish choughs; foliage at the top and sides of the shield.

"S' ELENE CORNVELL."—*Appended to a Charter by John Forrest, Provost of Linlithgow, with consent of Elene Cornwall and others, to Patrick Gibb, and Elizabeth Sym his spouse, of certain lands near Linlithgow, 18th June 1575. —Colston Charters.*

235. COTHRANE, PATRICK.

Barry of ten.

"S' PATRICH COTHRANE."—*Appended to a Sasine, dated 26th February 1596.— Edinburgh Charters.*

236. COUPER, ADAM, OF GOGAR.

A chevron, ermine (?) between three branches of laurel, with as many leaves each; the shield surrounded with pretty foliage.

"S' ADAM COUPER DE GOGAR."—*Appended to Charter by James Mowat, Chaplain of the Holy Cross in St. Giles', with consent of his patron, Adam Couper, of the piece of ground whereon the Chapel was built, to John Couper, son and heir-apparent of the said Adam Couper, 6th April 1608.—Edinburgh Charters.*

237. CRAB, WILLIAM, Burgess of Aberdeen.

Conché; a chevron between two fleurs-de-lis in chief, and a crab in base; crest, on a helmet, a cherub's head in profile; supporters, two swans (? herons) with wings expanded.

"s' WILELMI CRAB."—*Appended to Charter by William Crab to the Monastery of the Carmelites, of the yearly rent of some tenements in Aberdeen, 2d December 1499. —Marischal College, Aberdeen.*

238. CRAB, PAUL.

A chevron between two fleurs-de-lis in chief, and a crab in base.

"s' PAULUS CRAB."—*From the Collection of General Hutton, who only gives the reference, "Marischal College, 1310."*

239. CRAIG, JOHN, "Minister of Christ's Evangel."

Ermine; on a fess, a fleur-de-lis (?) between two crescents.

"s' M. JOHANNIS CRAIG."—*Appended to an Instrument dated 13th August 1584.— Edinburgh Charters.*

240. CRANEBORNE, THOMAS.

Conché; three cranes collared; crest, on helmet with mantlings, a crane's head and wings.

"s' THOME DE KRANEBROUN." Detached Seal.—*H.M. Record Office.*

241. CRANSTON, WILLIAM.

Three cranes passant.

"s' WILELMI DE CRANISTON."—*Appended to the Rebour of Simon Ward of Raynton, A.D. 1426.—Dean and Chapter of Durham.*

242. CRANSTON, THOMAS.

A crane's head, not on a shield; a monogram in front like the letter M.—*Appended to Treaty between King Henry Sixth and King James Second, 15th November 1449.—H.M. Record Office.*

243. CRANSTON, ANDREW.

A crane passant, not on a shield.

"s' ANDREE DE CRANIST." Detached Seal.—*H.M. Record Office.*

F

CRAWFORD, EARL OF. *See* Lindsay.

244. CRAWFORD, REGINALD.

A lion rampant, not on a shield.

"s' reginaldi de crawford."—*Appended to the same Homage as No. 48, A.D. 1292.*
—*H.M. Record Office.*

245. CRAWFORD, DEVORGILLA.

Oval shape, and of a well-executed design; a lady holding in
her right hand a shield bearing a fess chequé, and on her
left a falcon.

"sigill. derworgoul. d' crauford." Detached Seal.—
H.M. Record Office.

246. CRAWFORD, WILLIAM.

A fess between three mullets; a small cross at the top and
sides of the shield.

"s' willi de crawford." Detached Seal.—*H.M. Record Office.*

247. CRAWFORD, WILLIAM.

A fess, ermine; in dexter chief a mullet; and in the sinister a crescent.

"s' villelmi crawford."—*Appended to Sasine of Lochbank, 23d July 1546.*—*Edinburgh Charters.*

248. CRAWFORD, WILLIAM.

A fess, ermine, between a mullet in the dexter, a human heart in the sinister
chief, and a fleur-de-lis in base; foliage at the top and sides of the shield.

"s' villelmi crawford."—*Appended to a Sasine of ground at Leith, 24th January
1548.*—*Edinburgh Charters.*

249. CRAWFORD, AGNES.

A fess, ermine; in the sinister chief a mullet.

"s' agnes crawford."—*Appended to a Charter by Agnes Crawford, Lady of Les-
norcis, and heiress of Nicol Crawford of Oxingangis, to George Bog, servitour
to the Queen, of three acres of land in Linlithgow, 30th June 1549.*—*Colston
Charters.*

250. CRAWFORD, WILLIAM, of Lufnois.

Three stags' heads erased.

"s' guilelme craufurd."—Appended to Charter to John Murray of Blackbarony. 18th January 1587.—Blackbarony Charters.

251. CRICHTON, ROBERT, Provost of the Church of St. Giles, Edinburgh.

A lion rampant; slight foliage at the top and sides of the shield.

"s' dni roberti creitoun."—Appended to some Instrument as No. 217

252. CRICHTON, ROBERT, of Eliock.

Quarterly: first and fourth a lion rampant (Crichton); second and third three water bougets (Ross of Sanquhar).

"s' roberti creichtoun."—Appended to Procuratory of Resignation of the Lands and Barony of Cluny, Perthshire, by Mr. James Creichton, to James, Bishop of Dunkeld. Edinburgh, 20th June 1575. Communicated by John Stuart, Esq.

253. CRICHTON, JAMES, "The Admirable Crichton." Son of the above Robert Crichton.

The same as the preceding: foliage at the top and sides of the shield.

"s' m. jacobi creichttoune."—Appended to the same Instrument as the last.

254. CRISTISON, THOMAS.

The impression is much injured, but the charge appears to be an arm holding a heart palewise.

"s' thome cristison."—Appended to Charter to John Broan of Colston of an annual rent out of a tenement at Aberlady. Not dated, but probably about a.d. 1480 or 1490.—Colston Charters.

255. CULLAN, JOHN, of Knavane.

On a bend between two boars' heads couped, a cinquefoil inter two buckles, tongues erect.

"s' joannis cullan."—a.d. 1517.—Communicated by Lord Lindsay.

256. CUMIN, SIR WILLIAM. KNIGHT.

In bad condition. Equestrian design, knight riding to sinister; faint appearance of a garb on the shield. The inscription quite illegible.—*Appended to a Charter by Sir William Cumin, of one stone of wax yearly to St. Cuthbert at Durham, early in thirteenth century, certainly previous to* A.D. 1282.—*Dean and Chapter of Durham.*

257. CUMIN, JOHN, SON OF THE ABOVE SIR WILLIAM.

A fine seal; a shield bearing three garbs.

"s' DOM[INI JOHI]S CUMIN."—*Appended to Charter by John Cumin, with the consent of John, his son and heir, confirming the Charter of his predecessor, Sir William Cumin, of a stone of wax yearly.—Dean and Chapter of Durham.*

258. CUMIN, JOHN, SON OF THE EARL OF BUCHAN.

Three garbs.

" s' JOHIS COMYN FIL. COMIT. D. BUCHA."—*Appended to Submission of Competitors.* A.D. 1291.—*H.M. Record Office.*

259. CUMIN, EDMUND.

Three garbs; lizards around the shield.

" SIGILLUM EDMUNDI COMIN." Detached Seal.—*H.M. Record Office.*

260. CUMIN, FRANCIS.

A wheat sheaf, not on a shield.

" s' FRANCI CUM." Detached Seal.—*H.M. Record Office.*

261. CUMIN, JOHN.

An eagle, with wings expanded, not on a shield.

" s' JOHANNIS COMYN." Detached Seal. Rude workmanship.—*H.M. Record Office.*

262. CUMIN, WALTER.

Three garbs; over all a bend.

"s' WALTERI CUMIN." Detached Seal.—*H.M. Record Office.*

263. CUNINGHAM, WILLIAM, FOURTH EARL OF GLENCAIRN.

A shakefork.

" WILHELMI CUNIGHAM."—*Appended to Treaty for Marriage between King Edward VI. and Mary Queen of Scots, 1st July 1543.—H.M. Record Office.*

264. CUNINGHAM, ALEXANDER, Fifth Earl of Glencairn ("the good Earl").
Conché; a shakefork; crest on a helmet, a unicorn's head issuing from a coronet; supporters, two conies.

"ALEXANDER CONINGHAME COMES DE GLENCARNE."—*Appended to Charter by the Earl to Margaret Campbell, daughter of Colin Campbell of Glenurquhy, of the lands of Findlayson and others, 26th May 1574.—Communicated by W. Fraser, Esq.*

265. CUNINGHAM, JAMES, Grandson and Heir of the above Earl of Glencairn.
A shakefork; in dexter chief point a boar's head erased.

"S' JACOBI CUNINGHAM."—*Appended to a Tack, A.D. 1580.—Communicated by W. Fraser, Esq.*

266. CURRIE, SIMON.
A saltire, with a mullet in chief; foliage at the top and sides of the shield.

"S' SYMON DE CURRY."—*Appended to Charter by S. Currie and Helen Magil his spouse, dated 13th June 1588.— Edinburgh Charters.*

267. CURROR, WILLIAM, Burgess of Banff.
A fess between three mullets in chief and a hunting-horn stringed in base.

"S' VILELMI CURROR."— *Appended to a Charter dated 27th May 1567.— Edinburgh Charters.*

268. CUTHBERT, JOHN, of Auldcastlehill.
Apparently a merchant's mark; a star in the sinister chief and dexter base points.

"S' JOHANNIS CUTHBART."—*Appended to a Precept dated 30th April 1594.—Lovat Charters.*

269. DAIR, JOHN, Chaplain of the Altar of St. Ninian's (founded by Andrew Moubray) in the Church of St. Giles, Edinburgh.
A fess wattled; a rose and thistle ornament at the top and sides of the shield.

"S' JOANNIS DAIR CAPELLANI."—*Appended to a Confirmation of a sum of money to the Ministers for the use of the Poor of Edinburgh, 13th June 1654.— Edinburgh Charters.*

270. DALLAS, WILLIAM, of Budgate.
A heron's (?) head contourné; on a chief three mullets.

"SIG. VILELMI DALLYAS."—*Appended to Charter by William Dallas to J. Calder, precentor of Ross, 20th November 1540.—Communicated by Cosmo Innes, Esq.*

271. DALLAS, ALEXANDER.

A boar's head erased, contourné, and on a chief three mullets. Very rudely executed.

"s' ALLEXANDRI DOLAS."—*Appended to Charter by Alexander Dallas, son and apparent heir of William Dallas of Budzets, to Cathrine Campbell, "in her virginity," of the lands of Mylton of Candrayegelde, in the barony of Stratherne and sheriffdom of Inverness, at Elgin, 15th October 1540.—Cawdor Charters.*

272. DALZELL, ISABELLA.

A naked man, with arms extended.

"s' ISOBELLE DALZELL."—*Appended to Charter by Isabella Dalzell and her sister Alison, daughters of James Dalzell, burgess of Edinburgh, of part of the church lands of St. Cuthbert's, Edinburgh, 22d September 1529.—Edinburgh Charters.*

273. DANIEL, OR DANIELSTON, JOHN (DENNISTON).

A bend engrailed; in sinister chief a mullet.

"s' JOHANIS DE DANEELSTOUN."—*Appended to a Charter by John "Danyelston" of Leith, burgess of Edinburgh, of a piece of ground at Leith to the burgh of Edinburgh, 6th July 1439.—Edinburgh Charters.*

274. DEMPSTER, THOMAS, OF AUCHTIRLESS.

Quarterly: first and fourth, a lion rampant for Abernethy; second and third, a bar surmounted with a sword palewise, point in base, for Dempster.

"s' THOMAS DAMPSTAR D' AUTHIRLES."—*Appended to Charter, dated A.D. 1592.—Communicated by Lord Lindsay.*

275. DEMPSTER, JOHN, OF BALROWNY.

A rude example of art. The design, though heraldic, is not on a shield, unless the seal itself be taken to represent a circular one. Quarterly: first and fourth, a sword bendwise, point in base, surmounted by a fess; second and third, a lion rampant contourné for Abernethy.

"JOHNE DEMPSTER."—*Appended to Charter of the lands of Balrowny, in the parish of Menmuire, Forfarshire, to Patrick Livingstone, by John Dempster of Balrowny, A.D. 1607.—D. Laing, Esq.*

276. DENISTON, THOMAS.

A device, not on a shield, an arm issuing from the dexter side of the seal, holding a falcon by its jesses.

"s' THO[M]E DANASTOUN." Detached Seal.—*H.M. Record Office.*

277. DIXSON, ISABELLA, Wife of W. Nicolson.

Three mullets. Slight foliage surrounds the shield.

"S' ISABEL DYNCOUN."—*Appended to Reversion of one husband land in the town of Yester,* A.D. 1527.—*Tweeddale Charters.*

278. DOUGLAS, ...

A very curious design, not on a shield; a hand holding a figure very like the modern tourist's hand-bag.

"S' . . . ERO GRUMI DE DUGLAS." (?) Detached Seal.—*H.M. Record Office.*

279. DOUGLAS, WILLIAM.

A mullet, not on a shield.

"S' WILL. DE DUGLAS." Detached Seal.—*H.M. Record Office.*

280. DOUGLAS, WILLIAM.

On a chief three mullets; a lizard at the top and sides of the shield.

"S' DNI WILLELMI DE DUGLAS."—*Appended to Homage Deed,* 1296.—*H.M. Record Office.*

281. DOUGLAS, ARCHIBALD, Third Earl of, Lord of Galloway. *Plate I., Frontispiece, fig. 7.*

This seal is much broken; the parts remaining are, however, exceedingly fine and perfect.

Quarterly: first and fourth, a human heart, and on a chief three mullets for Douglas; second and third, a lion rampant, crowned, for Galloway; on an escutcheon surtout, three mullets for Murray of Bothwell. The dexter supporter—a savage—alone remains; the sinister has doubtless been the same. Above the shield, in the place usually occupied by the crest, is a singular kind of ornament, certainly not meant for a crest: it looks like a ribbon winding round three upright staves.

The background of the seal is filled up with foliage, and the quarters of the shield are prettily diapered.

"S' ARCHER[ALDI COMITIS DOUGLAS ET] DOMINI GALWYDIE."—*Appended to a Charter by the Earl of Douglas to John de Montgomery, Lord of Ardrossan, of the lands of Dunlop, in the barony of Cunynghame and shire of Ayr, 4th July* 1491.—*H.M. General Register House.*

This is an interesting seal, and perhaps the earliest Scottish example of the armorial ensigns of an heiress being carried on an escutcheon surtout; nor was the practice much earlier in England, the reign of Richard II. (1377-99) being given by Gwillim as the period of an early instance. Those given by Mr. Seton in his " *Law and Practice of Heraldry* " are some years later.

The Earl of Douglas having married Jean, daughter and sole heiress of Thomas Murray, Lord of Bothwell, became entitled to carry the coat of that family impaled with his own; or, as in this seal, surtout, that practice then beginning. His son, before his elevation to the Dukedom of Touraine, carried it in the third quarter of his shield, as in No. 242 of " *Descriptive Catalogue of Scottish Seals;*" but subsequently we find the Murray quarter omitted, as in the following example. It was, however, resumed by his successors, and retained its place on the Douglas escutcheon till the extinction of that great family.

282. DOUGLAS, ARCHIBALD, Fourth Earl of, Duke of Touraine and Earl of Longueville, Son of the above Archibald. *Plate* VI. *fig.* 9.

A savage man, holding in his right hand a club and a shield. Quarterly: first, three fleurs-de-lis, for Touraine; second, a human heart, and on a chief three mullets, for Douglas; third, a saltire and chief, for Annandale; fourth, a lion rampant, crowned, for Galloway; his left hand holds the helmet, from which issues the crest, a plume of feathers; background crusillée.

" S' ARCHIBALDI, DUCIS TURRENE, COMIT. DE DOUGLAS ET DE LONGUEVILLE, ETC.," A.D. 1421.—*Communicated by W. Fraser, Esq.*

283. DOUGLAS, MARGARET, Daughter of Robert III., Wife of the preceding Earl of Douglas.

A fine seal, but rather damaged. Quarterly, Touraine, Douglas, Annandale, and Galloway, as in the last, impaling Scotland; the shield supported by an angel, whose head and wings appear above.

" MERGARETE DUCISSE TURONIE COMITESSE DOUGLAS DNE GALVIDIE ET VALLIS ANADIE." —*Appended to a Charter by the Countess of the lands of Drumgewan to Gilbert Grierson, dated at Threave, 9th April A.D. 1425.—Communicated by W. Fraser, Esq.*

284. DOUGLAS, WILLIAM, Second Earl of Angus.

Quarterly: first and fourth Galloway; second and third Douglas, as before; on a surtout a bend bearing three mullets for . . . foliage at the top and sides of the shield.

" S' WILLELMI DNI DOUGLAS."—*Appended to Relaxation of Brockholes, etc. etc., to St. Ebbe, Coldingham, dated at Bonkyl, 31st May 1429.—Dean and Chapter of Durham.*

285. DOUGLAS, WILLIAM, Earl of Angus. Same person as the last.

> This is rather smaller, but differs in nothing from the last except in the inscription, and appearing to have a base bearing a cross, or some other charge. It is remarkable that the shield of William, ninth Earl of Angus (*vide* No. 255 of "*Descriptive Catalogue of Scottish Seals*"), has a base with a cross counter componé, a bearing which it is difficult to account for. The escutcheon surtout is probably the feudal arms of some lordship or territory held by the Earl of Angus at the time, but of which no record has been found to justify a more exact appropriation.
>
> " s' vilelmi douglas comitis de angus."—*Appended to a Receipt for 113 marks from William Drax, Prior of Coldingham, for Resignation of Brockholes, 4th June 1429.—Dean and Chapter of Durham.*

286. DOUGLAS, GEORGE, Master of Angus.

> A human heart, and a chief bearing three mullets; foliage at the top and sides of the shield.
>
> The usual inscription is expressed only by the initials s. g. d.—*Appended to Reversion by G. Douglas to Sir John Stirling of Keir, of the lands of Balcarne, 27th August 1527.—Crawford Charters.*

287. DOUGLAS, SIR GEORGE, Brother to Archibald, Sixth Earl of Angus.

> Quarterly: first and second, a lion rampant, for Angus and Galloway; third, three piles, for Brechin; fourth, a fess chequé, surmounted with a bend charged with three buckles, for Stuart of Boukil; on an escutcheon, surtout, Douglas, as before.
>
> " s' dni georgii douglas equitis auratis."—*Appended to Treaty for Marriage between King Edward VI. and Mary Queen of Scots, July 1543.—H.M. Record Office.*

288. DOUGLAS, SIR GEORGE, of Lochleven. The faithful adherent of Queen Mary, whom he assisted in her escape from Lochleven in 1567.

> A small signet; three piles, very inaccurately executed, having more the appearance of the partition line called indented,—there can be no question however that piles are meant, the external ones charged with a mullet; in base a crescent for difference; at the sides of the shield are the initials G. D.—*Appended to an Instrument dated A.D. 1560.—David Laing, Esq.*

289. DOUGLAS, CHRISTIAN, Wife of Alexander Lord Home.

Three piles, the external ones charged with a mullet.

"s' cristene dovglas."—*Appended to Charter by Alexander Lord Home and Christian Douglas his spouse, of the lands of West Fenton, to Alexander Home of North Berwick,* a.d. 1591.—*Home Charters.*

290. DOUGLAS, SIR ARCHIBALD, of Cavers, Knight, Sheriff of Roxburgh.

Quarterly : first and fourth, a human heart, and on a chief three mullets for Douglas : second and third, a bend between six cross-crosslets fitchée for Marr.

"s' archibaldi duglas de cav. mile."—*Appended to Renunciation by Archibald Douglas of Cavers to Sir Thomas Cranston of that Ilk, Knight, and William his son and heir-apparent, of writs and services in respect of lands of the Mains of Denholm, in the barony of Cavers and sheriffdom of Roxburgh, 9th February* 1465.—*R. Almack, Esq., Melford, Suffolk.*

291. DOUGLAS, JAMES, of Cavers, Sheriff of Roxburgh, Grandson of the above Sir Archibald Douglas.

Douglas, as before, without the Marr quarter.

"s' jacobi dovglas."—*Appended to a Charter by James Douglas of Cavers to Sir William Cranston, Knight, of the lands of Denholm, 31st October* 1512.—*R. Almack, Esq., Melford, Suffolk.*

292. DOUGLAS, SIR ARCHIBALD, of Spot, Knight.

A human heart, and on a chief three mullets.

"s' d. archebalde dovglas."—*Appended to Charter by Archibald Douglas of Spot to Sir Walter Dundas, of the lands of Bordland, co. Peebles, 12th June* 1616.—*Blackbarony Charters.*

293. DOUGLAS, JAMES, of Torthorald.

Quarterly ; in each quarter a cross ; over all a human heart.

"s' dni jacobi dovglas de torthorvald," a.d. 1617.—*W. Fraser, Esq.*

294. DOUGLAS, ALEXANDER.

Ermine ; a chief bearing a human heart between two mullets of six points ; foliage at the top and sides of the shield.

"s' alexandri dovglas."—*Appended to Sasine of five plough lands of Meikle Geddes, in the barony of Geddes and sheriffdom of Nairn, to Alexander Campbell of Flinismoir, 18th March* 1561.—*Cawdor Charters.*

295. DOUGLAS, JAMES, BURGESS OF ELGIN.

A fess between two mullets in chief and four ermine spots in base.

"s' JACOBI DOUGLAS."—*Appended to a Sasine dated January 1627.--Elgin Charters.*

296. DRIFFELD, RALPH.

A bicorporate leopard, or lion rampant, crowned, not on a shield. Curious example.

"s' RADULFI D' DRIFFELD."—*Early in 13th century.—Dean and Chapter of Durham.*

297. DRING, PATRICK.

A device, not on a shield, of an eagle, with wings expanded, standing on a tree.

"SIGILL. PATR. DRENG DE REIMGTUNE."—*Appended to Confirmation of the lands of Remington to St. Ebbe at Coldingham.—Dean and Chapter of Durham.*

298. DRUMMOND, JOHN, ELDEST SON OF LORD MADERTY.

Three bars wavy, in chief an eagle's head (?); above the shield a helmet, but no crest, supporters, two savages, each holding a club in the exterior hand.

"SIGILLUM JOANNIS [MAGI]STRI DE MADERTIE," A.D. 1610.—*David Laing, Esq.*

299. DRUMMOND, JOHN, DESIGNED DUKE OF MELFORT, SECOND SON OF JAMES, THIRD EARL OF PERTH; TREASURER DEPUTE AND SECRETARY OF STATE FOR SCOTLAND; CREATED EARL OF MELFORT AND VISCOUNT FORTH, A.D. 1686.

The arms of Scotland within a bordure composé. Crest on helmet, with mantling, a lion affronté issuing from a celestial crown, holding in the dexter paw a sword, and in the sinister a thistle. Supporters, two lions rampant, gorged with collars, bearing thistles stemmed and leaved.

Motto on a scroll beneath the shield, "ELI BONO SUM QUOD SUM."—*From a very pretty triangular steel seal in the British Museum.*

On one of the remaining sides is a knight in armour, bareheaded, riding to the dexter, sword in his right hand; in the upper part of the background are the arms of Scotland in a cartouche-shaped shield. Motto surrounding, "AB ORIGINE SACRA."

The third side is occupied by a shield, supported by a lion sejant, holding a sword and thistle, as in the crest, containing the combined initials (J. D., S. L.) of John Drummond and Sophia Lundie.

300. DRUMMOND, CHRISTIAN.

Three bars wavy; foliage at the top and sides of the shield.

"s' CRESTENE DRUMUND."—*Appended to an Instrument dated 9th April 1602.—Edinburgh Charters.*

301. DRUMMOND, DAVID, OF MACHANY.

Three bars wavy, over all an anchor.

"SIG. D. DAVID DRU[MOND] DE BORL. ET LEDMACHANY." Very coarse work.—*Communicated by the late Henry Drummond, Esq.*

302. DRUMMOND, JOHN, LORD.

Three bars wavy. Crest on helmet, a talbot's head; supporters, two savages; background foliated.

"s' JOHANNIS DNI DRUMUND." Detached Seal.—*H.M. Record Office.*

303. DRUMMOND, JOHN, LORD.

Three bars wavy.

"s' JHOES DNS DRUMMOND. Very rude seal. Detached Seal.—*H.M. Record Office.*

304. DRYNGETS, THOMAS, OR DRUNGERT.

A device, not on a shield, a hare leaping.

"s' THOME DE DRYNGERTS." (?)—*Appended to the same Homage as No. 48, A.D. 1292.—H.M. Record Office.*

305. DUDINGSTON, ANDREW, OF SOUTHOUSE.

A chevron between three mascles, and a chief bearing as many mullets; a slight foliage at the top and sides of the shield.

"s' ANDREE DUDINSTOUN."—*Appended to Charter by Andrew Dudingston of the lands of West Loch, etc., to John Murray of Blackbarony, 9th January A.D. 1579.—Blackbarony Charters.*

306. DUNBAR, GOSPATRICK, EARL, SECOND EARL ACCORDING TO THE PEERAGES.

Equestrian design, knight in chain-mail galloping to the sinister.

"SIGILLUM GOSPATRICII FRATRIS DELFIN."—*Appended to a Charter by Gospatrick, of the lands of Ederham and Nesbit to St. Cuthbert at Durham, A.D. 1150.—Dean and Chapter of Durham.*

307. DUNBAR, PATRICK, EARL, son of
   the above Earl Gospatrick.

   A knight on horseback, proceeding to the
      sinister, in chain-mail, conical hel-
      met, a sword in his right hand, resting
      on the arm; on his left arm a shield,
      on which is some appearance of a bor-
      dure, the umbo quite distinct. The
      inscription is unfortunately nearly
      lost, but has probably been " SIGIL-
      LUM PATRICII COMITIS LONIEE." (?)

308. COUNTER SEAL of the last.

   An antique gem, fine Greek or Roman
      work, a man milking a goat.

   " SIGILLU. ROBERLLI · E · II · · " (?)—Ap-
      pended to a Charter by Earl Patrick,
      confirming that of his father, of
      Ederham and Nesbit to St. Cuthbert
      at Durham, circa 1150.—Dean and
      Chapter of Durham.

   This counter seal is deserving of remark,
      not only as one of the earliest ex-
      amples of the use of such seals,
      but more particularly for its being
      ascribed to Robert, Bishop of St.
      Andrews, as will be afterwards mentioned.

The custom of using such counter seals continued for some centuries subsequently, but
   was not very prevalent, except in the case of some of the burghs, and the Great
   Seals of the kingdom, the latter having, from this period, or rather earlier, been
   continued in use uninterruptedly till the present time, but these are of quite a
   different character from those we are now considering, which were always of
   much less size than the seal, and were engraved gems of Greek or Roman
   work, set in a metal collar, on which the legend was cut, and were evidently in
   many instances intended also for ornaments pendant from the arm or neck.

This seal and counter of Earl Patrick are engraved in Anderson's "*Selectus Diplo-
   matum et Numismatum Scotiæ Thesaurus*," but not very accurately; and in the
   introduction by the learned Ruddiman, he states the counter seal to be that

of Robert, Bishop of St. Andrews, as giving his sanction; but this seems not
very probable. If it was necessary that the Bishop should confirm the deed
with his seal, it would surely have been done by appending it on a separate tag.
Ruddiman has probably been misled by the first words of the legend, which,
it must be admitted, appear to be " SIGILLU. ROBER . . .   ," Robert being
at that time Bishop of St. Andrews.

309. EDGAR, Son of Earl Gospatrick.   *Plate IV. fig. 4.*
> A fine and curious seal. A dragon, or other fabulous animal, of course not on a
> shield.
> " HOC EST SIGILLUM EDGARI FILII GOPATRICH COMITIS."—*Appended to a Charter by
> Edgar to the Abbot of St. Albans and Monastery of St. Oswyn of Tynemouth,
> confirming the grant of his father of the Church of Edulningeham, about A.D.
> 1150.—Dean and Chapter of Durham.*

310. DUNBAR, WALDEVE, EARL.
> Equestrian design, similar to the earlier seals.
> " SIGILLUM WALDEVE COMITIS."—*Appended to an Indenture or Covenant between the
> Earl and convent of Coldingham regarding the lands of Remington, circa 1170.
> —Dean and Chapter of Durham.*

311. DUNBAR, PATRICK, EARL, Son of Earl Patrick.
> Similar equestrian design.
> " SIGILL. PATRICII FILII COMITIS PATRICII."—*Appended to a Letter to his Father
> regarding the Donation of 170 marks silver from his Ville of Sweynwode, A.D.
> 1233.—Dean and Chapter of Durham.*

312. DUNBAR, WILLIAM, Son of Earl Patrick.
> A wyvern attacking a lion.
> " SIG. WILL. FIL. COMITIS DUNBAR."

313. COUNTER SEAL of the last.
> An antique gem. A warrior on horseback at full speed to the dexter, over a pros-
> trate figure.
> " SIG. WILL. FIL. COMITIS."—*Appended to a Charter by Earl William, of Sweynwode
> to St. Cuthbert's at Durham.—Dean and Chapter of Durham.*

314. DUNBAR, PATRICK, Earl Dunbar.

> A shield, bearing a lion rampant within a bordure charged with roses; foliage on the dexter side of the shield and a lizard on the sinister.
>
> " s' PATRICII COMITIS D' DUNBAR."—*Appended to Cyrograph of G. Prior of Colding-ham, regarding the stream of Akesudeburc.—Dean and Chapter of Durham.*

315. DUNBAR, PATRICK, Earl of March. *Plate I., Frontispiece, fig.* 1.

> A remarkably fine seal; an armed knight on horseback, a sword in his right hand, and on his left arm a shield bearing a lion rampant within a bordure charged with eight roses, which are repeated on the housings. Crest, a horse's head bridled, all to the sinister.
>
> " SIGILLUM PATRICII DE [DUNBAR COMI]TIS MARCHIE."

316. COUNTER SEAL of the last. *Plate I., Frontispiece, fig.* 2.

> Equally fine. A shield, as before, in centre of a rose tressure, or circular panel cusped. Same inscription as on the obverse.—*Appended to a Quitclaim by the Earl of March to the Prior of Coldingham, of an annual payment of ten shillings, due from Edreham, 24th May* 1367.—*Dean and Chapter of Durham.*

317. DUNBAR, GEORGE, Earl of March.

> A lion rampant, within a bordure charged with eleven roses; a label of three points.
>
> " SIGILLU. COMITIS LA MERCH."—"*Sir James Balfour says this seal is appended to ' a Confirmation by William, first Lord Settone, of a donation given be David de Anandia miles, Patricio de Halwick Guardiano fratrum de Hadingtone . . . of als many colles as they can burne for their awen usse, ex villa sua et Baronia de Trancnt,' 26th November* 1380."—*Hutton's* " *Sigilla,*" *p.* 112.

318. DUNBAR, JOHN, Earl of Moray.

> An armed knight on horseback, galloping to the dexter, a sword in his right hand, and on his left arm a shield bearing three cushions within the double tressure, which charges are also repeated on the housings.
>
> " SI[GILL. JOHANIS COM.] MORAY [DNI] VA[LIS] ANNAND."

319. COUNTER SEAL of the last.

> A shield, as in the last; above it are three mullets, and at the sides an ornament of foliage.
>
> " SIGILLUM COMITIS MORAVIE."—*Appended to* " *Charter of the lands of Collodonn, in Moray, by the Earl of Moray to Sir Alexander Seton.*" *Copied from a Collection by Sir James Balfour.—Harleian MSS.,* 4693, *British Museum.—Hutton's* " *Sigilla,*" *p.* 107.

320. DUNBAR, JOHN, Earl of Moray.

Three cushions within the royal tressure. Crest on a helmet, a stag's head. Supporters, two lions sejant regardant. Inscription not deciphered.—*Appended to a Charter by Hugh Fraser of Kynnel to William Chalmers, of the lands of Auchthendalyn, in Forfarshire; not dated, but probably about A.D. 1380.—Communicated by W. Fraser, Esq.*

321. MORAY, MARJORY, Countess of, Eldest Daughter of Robert II., Wife of the above John Dunbar, Earl of Moray.

An interesting little seal, but unfortunately much broken. The shield, however, is entire, and bears the Royal Arms of Scotland. All that remains of the inscription are the letters " oravie," being the latter part of the title " Comitissa Moravie."—*Appended to an Instrument of John Earl of Moray, given at Elgin, 1st May 1390.—Communicated by C. Innes, Esq.*

322. DUNBAR, SIR DAVID, of Cockburn, Knight.

A lion rampant, within a bordure, charged with eight mullets. Crest on helmet, with mantlings, a horse's head bridled, issuing from a coronet.

" s' david dunbar."—*Appended to Charter by Sir David Dunbar of Cockburn, etc., Knight, to the College of St. Salvator, 12th December 1452.—St. Salvator's College Charters.*

323. DUNBAR, JAMES, of Conze.

Three cushions; in fess point a mullet as a difference.

" s' jacobi dunbar [de conze]."—*Appended to Precept of Sasine by James Dunbar of Conze, to infeft John Rose and Marjory Dunbar, daughter of the said James Dunbar, in the lands of Lagygam, etc., 15th August 1526.—Cawdor Charters.*

324. DUNBAR, GAVIN, Archdeacon of St. Andrews.

Three cushions within the Royal Tressure; foliage around the shield.

" s' magri gavini dunbar archi di. sci andree."—*Appended to a Sasine dated 1536.—St. Andrews Charters.*

325. DUNBAR, ALEXANDER, Dean of Moray.

The same charges as in No. 323.

" s' alexandri dunbar de d."—*Appended to Tack by Alexander Dunbar, Dean of Moray, and one of the Senators of our King's College of Justice, with consent of the Bishop and Canons of the Cathedral Kirk of Moray, to John Campbell of Cawdor, of the teind-sheaves of the lands of Rait and Geddes, 1st September 1586.—Cawdor Charters.*

326. DUNDAS, ALEXANDER.

A lion rampant.

"S' ALEXANDRI DUNDAS."— *Appended to Retour of Alexander Cockburn of Newhall, as heir to his father, Alexander Cockburn, of eleven acres of land at Sandersdeen, Haddington, 30th August 1518.—Colston Charters.*

DUNFERMLINE, EARL OF. *Vide* SETON.

327. DURIE, HENRY, MERCHANT AND BURGESS OF MUSSELBURGH.

A chevron between three crescents; a mullet in fess point as a difference.

"S' [HENRICI] DURE."—*Appended to a Charter by Henry Durie to the Burgh of Musselburgh, 7th November 1579.—Musselburgh Charters.*

328. DUREME, SIR WILLIAM, KNIGHT.

On a bend three mullets.

"S' WILELMI [DUREME]."—*Melrose Charters.*

329. ECCLES, ADAM.

Device of a dexter hand above a chalice.

"SIGNUM ADE SACERDOTIS," *circa* 1170.—*Melrose Charters.*

330. ECHELEN, RALPH.

Oval shape, not on a shield; St. Michael vanquishing the dragon.

"S' RADULFI DE ECHELEN." Detached Seal.—*H.M. Record Office.*

331. EDINGTON, ADAM.

On a chief three birds (? martlets) passant.

"SIGILL. ADE DE EDINGTUN."—*Appended to a Document in possession of the Dean and Chapter of Durham.*

332. EDINGTON, GILBERT.

On a bend three mullets; in the sinister chief a bugle-horn stringed.

"S' GILBERTI DE EDINGTON."—*Appended to an Inquisition in the Court of Sir Alexander, Lord Home, Bailie of Coldingham, A.D. 1453.—Dean and Chapter of Durham.*

333. EDINGTON, RICHARD.

A chevron between three birds passant.

"S' RICAEDE DE EDINGTON."—*Appended to a Document regarding some rights of Common, circa 1450.—Dean and Chapter of Durham.*

H

334. EDWARD, GEORGE.

On a chevron, between two mullets in chief, and a holly-leaf and a buckle, palewise, in base, a rose pierced.

"s' GEORGII EDWARDI."—*Appended to Charter of two husband-lands in Nether Ayton to the Monastery of St. Ebbe's, Coldingham, A.D. 1441.—Dean and Chapter of Durham.*

EGLINGTON, EARL OF. *Vide* MONTGOMERIE.

335. EGLINTON, RALPH.

A device of a rabbit feeding.

s' RADULFI DE EGLLTUN."—*Appended to Homage Deed, A.D. 1295.— H.M. Record Office.*

336. EICHT, ROBERT.

A chief, charged on the sinister with a lion's (?) head erased.

" s' ROBERTI EICHT."—*Appended to Inquisition of the lands of Lumsden, finding Alexander Lumsden, heir to his brother, Thomas Lumsden, A.D. 1444.—Dean and Chapter of Durham.*

337. ELLOUS, ROBERT.

A lion rampant, debruised with a bend; in sinister chief a mullet; the shield surrounded by fine tracery.

" SIGILLUM ROBERTI ELLOUS."—*Appended to a Document dated A.D. 1357.—Dean and Chapter of Durham.*

338. ELPHINSTON, WILLIAM, RECTOR OF CLATT AND CANON OF ABERDEEN.

A chevron between three boars' heads and necks erased.

" s' VILIELMI ELPHINSTONE," A.D. 1505.—*From the Hutton Collection.*

339. ELPHINSTON, JAMES, LORD COUPAR, SECOND SON OF JAMES LORD BALMERINO.

Seal much damaged. The charges are evidently, on a chevron between three boars' heads and necks erased, a human heart. Crest on full-faced helmet, with mantling, a stag's head. Supporters, two winged stags; the only instance of such supporters in Scotland. A ribbon above the helmet, containing only a few letters of the inscription. The inscription surrounding, though imperfect, is evidently " SIGILLUM JACOBI ELPHINSTON DOMINI CUPRI."—*Appended to Charter by Lord Coupar, with consent of his brother John Lord Balmerino, in favour of Margaret Haliburton, daughter of Sir James Haliburton of Piteur, 13th December 1620.—Communicated by W. Fraser, Esq.*

340. ERSKINE, ALEXANDER, Lord.

Couché; a pale. Crest on helmet, a swan's neck and wings; foliage in background
" s' ALEXANDRI DNI ERSKYN." Detached Seal.—*H.M. Record Office.*

341. ERSKINE, SIR ROBERT, sixth Lord, Chamberlain of Scotland.

Couché; a pale. Crest on a helmet, with mantling, a boar's head and neck.
" s' ROBERTI DE HIRSKYNE."—*Appended to an Instrument relating to the Ransom of
King David II., 5th October 1357.—H.M. Record Office.*

342. ERSKINE, SIR JOHN, of Dun.

Couché; on a pale, a cross-crosslet, fitchée. Crest on helmet,
with mantling, a griffin's head issuing from a coronet.
" s' JOHANNIS DE ERSKYN DN. DUN."—*Appended to Charter
of the lands of Carkary, in Forfarshire, to Walter
Ogilvy, A.D. 1400.—Communicated by W. Fraser, Esq.*

343. ERSKINE, JOHN.

Quarterly: first and fourth, a pale; second and third, a bend.—*Appended to Ratification of Truce for one year, between King Henry VIII. and King James V.,
7th October A.D. 1518.—H.M. Record Office.*

344. ERSKINE, JOHN, Lord, Earl of Marr, etc. etc., K.G. *Plate V. fig. 6.*

A fine seal. Quarterly: first and fourth, a bend between six cross-crosslets, for
Marr; second and third, a pale, for Erskine; the shield surrounded with the
Garter and motto of the Order. Crest on helmet, with mantlings above a
coronet, a dexter hand couped, holding a sword fessways. Supporters, two
griffins. Motto on a ribbon below, " JE PENSE PLUS."
" s' JOHANIS COMITIS DE MAR DOMINI ERSKYNE ET GARVIOCH."—*Appended to Charter
of John Earl of Marr, dated at Holyrood House, 24th December 1623, in
possession of Francis G. Fraser, Esq. of Findrack, Aberdeenshire.*

345. ERSKINE, SIR CHARLES, of Cambo, Baronet, Lyon King-at-Arms.

Quarterly: first and fourth, a royal crown within the tressure of Scotland, being the
coat of augmentation granted to Thomas, the first Earl of Kelly; second and
third, a pale charged with a crescent for difference, Sir Charles being the
second son of Alexander, third Earl of Kelly; on an escutcheon surtout, the
badge of the Knights of Nova Scotia. Crest on full faced helmet, with
mantling, a demi-lion rampant.

" SIGILLUM OFFICII LEONI ANNO DOMINI 1663."—*Appended to a Warrant, dated 23d June 1668, to Sir Jerome Spens, Rothesay Herald, to cite and charge Robert Dunbar, younger of Burgie, to appear before the Lords of Privy Council on 30th July 1668.—Edward Dunbar Dunbar, Esq. of Sea Park, Forres.*

The proper official seal of the Herald Office, as now used, will be found described in the class of official seals. It was not adopted till 1673. See Seton's " *Law and Practice of Heraldry,*" p. 32.

346. ERSKINE, JOHN.

Couché ; on a pale, a cross crosslet, fitchée. Crest on helmet and mantling, a griffin's head erased.

" SIGILLUM JOHANNIS ERSKINE."—*Appended to an Instrument dated 1532.—Crawford Charters.*

347. ETAL, JOHN.

A rose (?).

" S' JOHANNIS D. ETAL."—*Appended to Charter by John Etal and Alicia Pynchist his spouse, of the rent of a tenement in Aberdeen to the Convent of the Carmelites, 4th July 1399.—Marischal College, Aberdeen.*

348. EVIOT, JOHN, OF BALOWSE (BALHOUSIE).

On a bend five bezants, or other roundles.

" S' JOHANNIS EVIOT."—*Appended to Charter by John Eviot to his Son Robert Eviot, of all the barony of Balhousie, A.D. 1448.—St. Salvator's College Charters.*

349. FAIRFOWL, OR FAIRFUL, JOHN.

A mullet, and in the dexter chief a saltire couped.

" S' JOHANNIS FAIRFUL."—*Appended to an Ecclesiastical Instrument dated 4th April 1480, in possession of David Laing, Esq.*

350. FAIRHOLM, ... SECOND SON OF FAIRHOLM OF CRAIGIEHALL.

An anchor, in dexter chief a crescent as a difference. Crest on a helmet, with mantling, a dove standing, holding a branch of laurel in its beak.

Motto on ribbon above, " MELIORA SPERO."—*From a pretty steel signet, the work of the early part of the seventeenth century, in the possession of W. Fairholm, Esq. of Chapel.*

351. FALCONER, ROBERT.

Device of an eagle preying on a bird.

" S' ROBTI FALCONER."—*Appended to Homage Deed, 1295.—H.M. Record Office.*

SCOTTISH SEALS. 61

352. FARQUHAR, ALEXANDER.

A stag's head erased, contourné; in base two roses, curiously executed, in outline
only.

" s' ALEXANDRI M. FARQUHAR."—*Appended to Charter of Reversion by Alexander
M'Farchyr M'Couche M'Anckryl to John Campbell of Cawdor, of the lands of
Dunmaglass, in the barony of Cawdor and sheriffdom of Inverness, 24th
March 1580.—Cawdor Charters.*

353. FARQUHAR, OR FERCHARD, DUNCAN.

A stag's attires, with a mullet between them.

" s' DUNCANI MAK FERCHEIR (?) DUNCANSON FERCHARD."—
*Appended to Charter of Reversion by Ferchard Dun-
canson, with consent of Duncan Maklen, in Dunmaglass, his father, to Sir John Campbell of Cawdor, of
the " sunny half " of the lands of Dunmaglass, in the
barony of Cawdor and sheriffdom of Nairn; dated
Perth, 2d December 1535.—Cawdor Charters.*

354. FARRER, JOHN.

Very imperfect. A bend sinister between four crescents.
" s' JOANNIS [FA]RAR," A.D. 1590.—*Aldbar Charters.*

355. FASSLENE, OR FOSSLAN, WALTER. *Plate VII. fig. 8.*

A saltire, cantoned with four roses; the shield in centre of tracery.

" s' VALTERI D. [F]OSLEN."—*Appended to Charter by " Walterus de Ffoslone duo de
Levenox," to Walter Buchanan, " duo ciusdem," and Margaret his spouse, of
the fourth part of the lands of Cambren: not dated, probably about A.D. 1440.
—Communicated by W. Fraser, Esq.*

356. FAULO, OR FAULAU, GEORGE, PROVOST OF EDINBURGH.

On a chevron, between three crosses fleury, a mullet; the shield supported by a
dragon with wings expanded.

" s' GEORGIUS DE FAULAU." (?)—*Appended to an extract from the Records of Edinburgh of the Commission given by Charles VII., King of France, to John Earl
of Buchan, appointing him Captain of 150 Scotsmen, to be the King's Guard, and
granting him a yearly pension of three thousand crowns, as a reward of his
services; dated at Paris 9th December 1421, registered in the public Records
of the City of Edinburgh, and extracted thence 16th February 1451, by Lady
Margaret Stuart, daughter and heiress of the late John Earl of Buchan, and
spouse of George Lord Seton.—Hutton's " Sigilla," p. 162.*

357. FAUSIDE, ROGER.

    A crane passant, within a bordure engrailed; the shield in centre of tracery.

    "SIGILLUM ROGERI DE FAUSYDE."—*Appended to a Charter by Anabell de Cogane, of some lands to the Priory of Coldingham, A.D. 1326.—Dean and Chapter of Durham.*

358. FEMOR, JOHN.

    Oval shape. A flower or wheel ornament.

    "S' JOHES DE FEMOR." Detached Seal.—*H.M. Record Office.*

359. FENTON, WILLIAM, OF FENTON.

    A chevron between three crescents; the shield surrounded by tracery.

    "S' WILLMI DE FENTON."—*Appended to an Inquest regarding the rights of the citizens of Brechin to hold a fair on Sundays, etc., 21st July 1450.—Brechin Charters.*

360. FIFE, DUNCAN, EARL OF.

    A mere fragment of a seal, exhibiting an equestrian figure, of a similar design to No. 334 of "*Descriptive Catalogue of Scottish Seals.*"—*Appended to a beautifully written Charter by Duncan (twelfth?) Earl of Fife, of the lands of Rothmelry (Rumeldrie) to John Monypeny, c. 1320.—George Seton, Esq. The lands of Rumeldrie subsequently belonged to the Setons of Cariston.*

361. FIFE, THOMAS, BAILIE OF ABERDEEN.

    Couché; a lion rampant. Crest on helmet, a swan's neck.

    "S' THOME DE FYF," A.D. 1486.—*From the Hutton Collection.*

362. FINDON, PHILIP.

    A device of a man's head in front, above it a crescent.

    "S' PHILIPPI DE FINDUS."—*Appended to Homage Deed, 20th August 1295.—H.M. Record Office.*

363. FISHER, WILLIAM, BURGESS OF EDINBURGH.

    Three fishes—salmon (?)—naiant, in pale; foliage around the shield.

    "S' VILLELMI FYCHER."—*Appended to a Charter dated 7th August 1567.—Edinburgh Charters.*

364. FLANDERS, WILLIAM.

    A fess surmounted by a bend.

    "S' WILL. FLANDRIENSIS." Detached Seal.—*H.M. Record Office.*

365. FLEMING, ALAN.

A star or wheel ornament.

" s' ALANI FLEMING." Detached Seal.—*H.M. Record Office.*

366. FLEMING, MALCOLM.

A chevron, within a double tressure flowered and counter-flowered; the shield surrounded by pointed tracery, in the spaces of which are a stag, a bird, and a dog.

" s' MALCOLMI FLEMING."—*Appended to Appointment of Plenipotentiaries for Ransom of King David II., 26th September 1357.—H.M. Record Office.*

367. FLEMING, JOHN, EARL OF WIGTON, LORD OF CUMBERNAULD.

Much injured, but has been a fine large seal. Quarterly: first and fourth, a chevron between three lions' heads erased, within the royal tressure, for Fleming; second and third, six cinquefoils, two, two, and two for Fraser. Crest on a coronet (like a celestial crown), above a full-faced helmet with mantlings, a stag's head erased; motto on a ribbon issuing from behind the helmet, probably " LET THE DEED SHAW," which has been the family motto for many generations, but in this instance there seems to be some additional word. Supporters, two stags.

" SIGILLUM JOHIS COMITIS DE VIGTOUNE ET DOM. FLEMYNG."—*Appended to Precept of Clare Constat in favour of John Broun, lawful brother and nearest heir of the late James Broun, to a bovate of land of Chabaurslands, 27th July 1644.—Colston Charters.*

368. FLEMING, WILLIAM.

Oval shape. The decollation of St. John the Baptist; in base a monk praying.

" s' WILLI FLEM . . . N CAS. CLICI." Detached Seal.—*H.M. Record Office.*

369. FODGAY (FOTHERINGHAM?), ROGER.

The Agnus Dei.

" s' ROGERI DE FODGAU."—*Appended to the same Deed of Homage as No. 48 A.D. 1292.—H.M. Record Office.*

370. FORBES, HENRY, BAILIE OF ABERDEEN.

Three bears' heads muzzled and couped; in centre point a crescent as a difference.

" s' HENRICI FORBES," A.D. 1594.—*P. Chalmers, Esq.*

371. FORBES, ROBERT, OF RYRES.

Three bears' heads couped.

" s' ROBERTI FORBES DE RERE," A.D. 1616. Detached Seal.—*Charles Baxter, Esq.*

372. FORRESTER, ARCHIBALD, of Corstorphine.

Rather indistinct. Three bugle-horns stringed; crest on helmet with mantlings, a dog's head.

" s' ARCHIBALDI FORSTR DXS DE CORSTON." Detached Seal.—*H.M. General Register House.*

This is probably the seal mentioned by Nisbet, vol. i. p. 423, as that of Archibald Forrester of Corstorphine, A.D. 1482, although the supporters are not visible.

373. FORRESTER, JAMES, of Corstorphine.

Three hunting-horns stringed. Crest on helmet with mantlings, a dog's head erased.

" s' JACOBI FORESTER DE CORSTORFYN."—*Appended to Charter of ground at Leith, called the Links, to W. Black, 13th July* 1547.— *Edinburgh Charters.*

374. FORRESTER, JOHN, Baron of Liberton.

A hunting-horn, not on a shield.—*Appended to " Concordia " by the Commissioners for governing the Marches, 12th July* 1429.—*H.M. Record Office.*

375. FORRESTER, WILLIAM.

Two hunting-horns, stringed, in chief, and a mullet in base.

" s' WILMI FORSTAR."—*Appended to a Sasine of a tenement in Leith in favour of James Makyson, A.D.* 1527.— *Edinburgh Charters.*

376. FOTHERINGHAM, HUGH.

A stag's head embossed, not on a shield.

" s' HUGONIS DE FODERIGAY."—*Appended to the same Homage as No.* 48, A.D. 1292. —*H.M. Record Office.*

377. FOTHERINGHAM, THOMAS.

Ermine. Two bars. Very rude work.

" s' TOME DE FOTHIRINGAME."—*Appended to Precept of Clare Constat by Alexander Master of Crawford, infefting Sir David Lindsay, his Cousin, in the lands of Cairncross, 25th September* 1475.— *Crawford Charters.*

378. FOTHERINGHAM, THOMAS, of Powrie Wester.

Four bars ermine; foliage at top of the shield.

" s' THO. FOTHRINGHAM DE POWRIE."—*Appended to Resignation by Thomas Fotheringham, and his son and heir, of lands in Perthshire, to Sir Duncan Campbell of Glenurquhay, 2d February* 1593.—*Breadalbane Charters.*

379. FOTHERINGHAM, DAVID, of Powrie.

Ermine; a bend.

"s' DAVID DE FOTHERINGHAM."—*Appended to an Inquisition dated* A.D. 1450.—*Brechin Charters.*

380. FOUDAS, LANCELOT.

A fess between three lozenges.

"s' LACELAT FOUDAS."—*Appended to Precept of Sasine in favour of James Innes of "yat ilk," A.D. 1468.—Communicated by Lord Lindsay.*

381. FOULAR, WILLIAM.

A rose between three cross-crosslets, fitchée.

"s' MAGISTRI WILLI FOULAR."—*Original brass matrix, in possession of John Stirling of Kippendavie, Esq., found in the old mansion of Viscount Strathallan, at Dunblane.*

382. FOULIS, WILLIAM.

A wheat sheaf, not on a shield.—*Appended to " Concordia," by the Commissioners for governing the Marches, 12th July 1429.—H.M. Record Office.*

383. FOULIS, SIR PHILIP, of Dunbar.

A mullet pierced, not on a shield. Letters between the points.—*Appended to the same document as the last.*

384. FOULLIS, HENRY, Chaplain.

A chevron between two mascles in chief and a mullet in base.

"SIG. HENRICUS FOLLIS."—*Appended to a Charter by Henry Foullis, with consent of his brother Thomas, to John Bathgryasi (?), of lands or tenements in St. Andrews, 20th April 1450.—University of St. Andrews Charters.*

385. FRANCIS, RICHARD.

A lion rampant, not on a shield.

"s' RICARDI FRANSEVSON."—*Appended to Quitclaim by Richard Francis, superior of Ayton, of a Messuage at Ayton, to St. Cuthbert's at Durham.—Dean and Chapter of Durham.*

386. FRANCIS, WILLIAM.

Device, two hands clasped; foliated background.

". . . WIDE VIBE ME DOMINE." (?) A poor seal.—*Appended to the Homage of " Will. Francis, Chevalier," 1296. —H.M. Record Office.*

I

387. FRASER, WILLIAM, "le fitz jadys Mons. Alisandre Fraser."
A label of three points, each charged with two " Frasiers."
"s' Wilelmi Fraser.—*Appended to Homage Deed*, 1295.—*H.M. Record Office.*

This is strictly heraldic, though not on a shield,—merely a fanciful mode of marking cadency. The six " Frasiers" constitute the well-known coat of Fraser, and are here carried by the son on a label, the proper cadency. It may be mentioned that this seal forms one of the many pleasing illustrations of Mr. Seton's valuable work, and is the only one to which even a qualified censure can be applied. The artist having most unaccountably drawn it upside-down, has made it quite incomprehensible to most readers.

388. FRASER, HUGH, first Lord Lovat.
Quarterly: first and fourth, three cinquefoils for Fraser; second and third, three crowns for Lovat (?). Crest on helmet, a stag's head. Supporters, two lions sejant.
" hugonis fraser dni lovat," A.D. 1431. Very imperfect.—*Kilravock Charters.*

389. FRASER, HUGH, second Lord Lovat, son of the above Hugh Lord Lovat.
Couché. Quarterly: first and fourth, Fraser; second and third, Lovat, as in the last. Crest on a helmet, with mantlings, a stag's head.
"s' hugonis dni lovat."—*Appended to Contract between William, Thane of Cawdor, and Alexander Fraser, Lord of Philorth, son of Hugh Lord Lovat, relative to the marriage of Marjorie Cawdor, daughter of the said William,* 1487.—*Cawdor Charters.*

390. FRASER, THOMAS, third Lord Lovat, son of the above Hugh Lord Lovat.
A crown between three cinquefoils.
"s' thome fresser."—*Appended to Precept by Henry Douglas of Culbriny to Thomas Lord Lovat, 15th October* 1509.—*Lovat Charters.*

391. FRASER, HUGH, fourth Lord Lovat, son of the above Thomas Lord Lovat.
Couché. Quarterly, Fraser and Lovat, as before. Crest the same as in No. 389.
's' hugonis dni de fresil."—*Appended to Charter by Lord Lovat to Hugh Symson of the lands of Erchet, 25th October* 1541.—*Lovat Charters.*

392. FRASER, HUGH, sixth Lord Lovat, Grandson of the fourth Lord.

Quarterly: first and fourth, three crowns for Lovat (?); second and third, three cinquefoils for Fraser. Surrounding the shield are the initials H · F · L · L ·

"S' HUGONIS FRESER."—*Appended to Charter by Lord Lovat to Alan M'Ranald, and Margaret Fraser his spouse, of the lands of Ester Leis, 17th December 1576.— Lovat Charters.*

393. FRASER, SIMON, seventh Lord Lovat, Son of the above Hugh Lord Lovat.

Much damaged, but evidently Fraser and Lovat, quarterly, as before.

"SIMONIS DOMINI FRASER DE LOWET."—*Appended to a Charter by Lord Lovat of the lands of Clonbackie to Hugh Fraser of Culbackie, 20th May 1615.—Lovat Charters.*

394. FRASER, HUGH, of Guschans.

Quarterly: first and fourth, three arched crowns, Lovat (?); second and third, Fraser, as before; foliage of the rose and thistle at the top and sides of the shield.

"S' HUGONIS FRESER DE GUSCHAN," A.D. 1582.—*Gairloch Charters.*

395. FRASER, SIR PETER, Bart., of Dores, the last of the direct line of Sir Alexander Fraser, Great Chamberlain of Scotland, A.D. 1396, now represented in the male line by Francis G. Fraser, Esq. of Findrack, Aberdeenshire.

A small signet. Three cinquefoils. Crest, a stag's head. Supporters, two stags — *From a Letter of Sir Peter Fraser's to Professor Fraser of King's College, Aberdeen, dated 8th October 1705. In possession of W. N. Fraser, Esq.*

396. FRENDRAUCH, DUNCAN.

Three wolves' (?) heads; the shield in centre of very pretty tracery.

"S' MAL...." (?)—*Appended to Homage Deed of "Duncan de Frendrauch, Che."* 1296.—*H.M. Record Office.*

397. FREELAND, ROBERT,

A fess chequé between two boars' heads and necks couped; foliage at the top and sides of the shield.

"S' ROBERTI FREELAND."—*Communicated by J. C. Roger, Esq., F.S.A. Scot.*

398. FUISNER, JOHN.

> Device of a wyvern; a star in front.
> " s' johis le fuisner."   Detached Seal.—*H.M. Record Office.*

399. FULLARTON, JOHN, of Fullarton.

> On a fess, three ermine spots; foliage around the shield. (Nisbet gives three otters heads erased as the bearings of Fullarton of that Ilk)
> " s' johanis fullartoun."—*Appended to Charter by David Fullarton, son and apparent heir of John Fullarton of Fullarton, to David Blair of Adamson, of the lands of Corsby and Craiglaw, co. Ayr, 31st May 1549.—David Laing, Esq.*

400. FULLARTON, DAVID, son of the above John Fullarton.

> Precisely the same charges as in the last.
> " s' david fullartoun."—*Appended to the same Instrument as the last.*

401. FULLARTON, WILLIAM, of Ardo.

> On a fess, between three boars' heads and necks erased, a crescent inter two mullets; foliage at the top and sides of the shield.
> " s' villielmi fulertoun."—*Appended to Reversion of the lands of Cultraw, a.d. 1575.—Communicated by the late P. Chalmers, Esq.*

402. FYTCHET, RICHARD.

> A chevron between three pea-pods (fitches); the shield surmounted by tracery.
> " s' ricardi fytchet."—*Appended to Charter by Patrick Sharp to John, son of Edmund " Medica" of Aberdeen, of some lands in Rubislaw, 15th January 1405.—Marischal College, Aberdeen.*

403. GARDINER, HENRY.

> A flower, not on a shield.
> " s' henrici d. gardino."   Detached Seal.—*H.M. Record Office.*

404. GARDYN, DAVID.

> Per fess, in chief a boar's head and neck erased, between two mullets; in base two cross crosslets, fitchée.
> " s' david gard[yn]," a.d. 1565.—*Communicated by Cosmo Innes, Esq.*

405. GARIOCH, ADAM.

A fleur-de-lis, not on a shield.

"S' ADAM DE GARVIAUTH." Detached Seal.—*H.M. Record Office.*

406. GARIOCH, ANDREW.

Oval shape. Under a Gothic canopy a figure of St. Andrew on the cross, and at each side a figure engaged in tying his arms to the upper limbs; in the lower part of the seal a monk praying.

"S' ANDREE DE GARVIACH CLICI." Detached Seal, probably thirteenth century work.—*H.M. General Register House.*

407. GARIOCH, LAURENCE, ALDERMAN OF ABERDEEN.

A very pretty seal. A chevron between three acorns slipped; the shield in centre of pointed tracery.

"S' LAURENCI D. GARWOC," *about* A.D. 1380.—*Glammis Charters.*

408. GAWENSONE, ALEXANDER.

A saltire between a mullet in chief, and a spear-head in base; foliage at the top and sides of the shield.

"S' ALEXANDRI GAWIESON."—*Appended to a Charter by Alexander Gawensone to his brother-german, Donald Gawensone, of a tenement of land in the burgh of Nairn, 1st March 1563.—Cawdor Charters.*

409. GIBB, JAMES, OF CARRUBERS.

A dexter hand couped, issuing from the sinister, holding a sword in pale, the upper part broken and falling to the sinister; in the dexter chief point a spur fesswise.

"S' JACOBUS GEIB OF CARRUBER." Detached Seal.—*H.M. General Register House.*

410. GIBSON, DAVID, VICAR OF COLMANEL, AYRSHIRE.

A fess between three fleurs-de-lis in chief, and a chalice, or covered cup, in base.

"DAVIDE GIBSONE."—*Appended to a Lease of the Vicarage to Gilbert Kennedy, in consideration of certain sums paid by him and his friends for repairing the University, etc., 24th July 1552.—Glasgow College Charters.*

411. GIBSON, DAVID, Canon of the Metropolitan Church of Glasgow.

Quarterly: first and fourth, two keys in saltire; second, gyronny of eight; third, a saltire. The impression is much damaged, and accordingly the blazon of the third and fourth quarters may not be quite correct.

"S' DAVID GIBSONE."—*Appended to an Ecclesiastical Instrument dated 5th April 1557.—David Laing, Esq.*

412. GIBSON . . . . . of Pentland.

Three keys fesswise in pale, wards downwards, turned to the sinister (which must be a mistake of the engraver), within a bordure vairé. Crest on a helmet, with mantling, a pelican feeding her young.

" PANDITE CŒLESTES PORTÆ."—*Steel matrix, early eighteenth century work, in possession of J. T. Gibson-Craig, Esq.*

Gibson of Pentland now carries this coat without the bordure, the main line of Durie being extinct.

413. GIFFORD, ADAM, Bailie of Aberdeen.

Barry of six argent and ermine; in sinister chief point a boar's head couped.

"S' AD[AM]E GIFFARD."—*Appended to Charter by Patrick Sharp to William Russ, Chaplain, 2d October 1408.—Marischal College, Aberdeen.*

414. GIFFORD, JAMES, of Sheriffhall.

Barry ermine and argent (no doubt meant for three bars ermine).

"SIGILLUM JACOBI GIFFURD DE SHIRIFHALL," A.D. 1405.—*Hutton's "Sigilla," p. 186.*

415. GILBERT, MICHAEL, one of the Bailies of Edinburgh.

On a chevron, between three trefoils slipped, a fleur-de-lis; foliage at the top and sides of the shield.

"S' MICHAELIS E. GILBERT."—*Appended to a Sasine, 21st September 1588.—Edinburgh Charters.*

416. GLADSTONE, HERBERT.

A flower or wheel ornament.

"S' HERB. D. GLEDSTAN." Detached Seal.—*H.M. Record Office.*

417. GLADSTONE, JOHN, of Cocklaw.

A lion rampant.

"S' JOHANNES GLADSTON'S."—*Appended to "Tak" of the lands of Marchedike, 12th November 1509.—Philiphaugh Charters.*

418. GLENCUK, MORGAN (?).

Semé of cross-crosslets, fitchée : a bend bearing three charges, too indistinct to specify.
" s' MORGUNIDE GLENC . . K."(?)   Detached Seal.—*H.M. Record Office.*

419. GOCELIN, LAWRENCE.

Au estoile and crescent, not on a shield.
" s' LAURENCI GOSSLIN."—*Appended to Charter by Lawrence, son of William, son of
" Gocelin," to St. Ebbe, Coldingham, of a bovate of land at Prendergast.—
Dean and Chapter of Durham.*

420. GOSLIN (SUBSEQUENTLY STEVENSON), JOHN, SON OF ARCHIBALD.

A crescent and au estoile, not on a shield.
" SIGILL. JOHANNIS DE GOSLIN."—*Dean and Chapter of Durham.*

421. GORDON, SIR ALEXANDER, KNIGHT, CREATED EARL OF HUNTLY, A.D. 1449.

Very imperfect, unfortunately, but evidently the quartered coat of Seton and Gordon.
   Crest on helmet, a boar's head and neck between two demi-vols issuing from a
   coronet ; the background filled up with foliage and two trees growing from a
   mount.
" s' ALEXANDRI SETON DNI DE GORDON."—*This seal was used by Crichton, the Lord
   Chancellor, in absence of his own, as well as by Sir Alexander Gordon, for
   himself.—Appended to the same Instrument as No.* 11.

422. GORDON, ALEXANDER, LORD, AFTERWARDS THIRD EARL OF HUNTLY.   *Plate
VIII. fig.* 1.

Couché.   Quarterly : first, three boars' heads for Gordon ; second, three lions' heads
   erased for Badenoch ; third, three cinquefoils for Fraser ; fourth, three crescents
   within the royal tressure for Seton ; over all a label of three points.   Crest on
   helmet, a stag's head.   Supporters, two greyhounds.
" s' ALEXANDRI DNI GORDONE."—*Appended to Precept of Sasine by Alexander Lord
   Gordon, Master of Huntly, in favour of Alexander Lord Home, 8th November
   1492.—Home Charters.*

423. GORDON, ALEXANDER, LORD.   SAME PERSON AS THE LAST.

The same bearings exactly as the last, but without crest, supporters, or ornaments
   of any kind.
" SIGILLUM ALEXANDRI GORDON."—*Appended to Precept of Sasine in favour of Lord
   Home, 24th July 1498.—Home Charters.*

424. GORDON, ALEXANDER, third Earl of Huntly. Same person as the last.
    Couché. Quarterly, as before, but without the label, the Seton coat occupying the *third* quarter, and with crest, etc. as in No. 422.
      " s' alexandri comitis de huntlie."—*Appended to Precept of Sasine in favour of Alexander Lord Home. 14th May 1503.—Home Charters.*

425. GORDON, ALEXANDER, third Earl of Huntly. Same person as the last.
    In every respect the same as the last, but of a larger size. Detached Seal.—*Home Charters.*

426. GORDON, ALEXANDER, third Earl of Huntly. Same person as the last.
    Quarterly, as before; crest, etc., the same.
      " s' alexandri comitis huntlie."— *A lead or pewter matrix, found in a moor in the south of Scotland, now in possession of the Duke of Richmond. Interesting particulars regarding this seal will be found in the Archæological Journal, vol. x. p. 335.*

427. GORDON, ADAM.
    A lion's head to sinister, not on a shield.
      " s' ade de gordwne." Detached Seal.—*H.M. Record Office.*

428. GORDON (?), HUGH.
    Oval shape. A patriarchal cross.
      " s' hugonis de gordonede." (?) Detached Seal.—*H.M. Record Office.*

429. GORDON, JANET, Lady Danachtan.
    A bend indented between four bears' heads couped, and two cross crosslets, fitchée; foliage at the top and sides of the shield.
      " s' janete gordon dne de dunach . . ."—*Appended to Reversion of the lands of Daltullich by the Lady of Danachtan, to Hugh Lord Lovat, 9th November 1528.—Lovat Charters.*

430. GORDON, JANET, Wife of James Ogilvy, of Cardell.

A bend dancette between three boars' heads erased.

" S' JONETE GORDONE."—*Appended to a Charter by Janet Gordon, with consent of James Ogilvy of Cardell, her spouse, of the lands of Dalcorse, to James Ogilvy, their son and heir, 27th October 1558.—Lovat Charters.*

431. GORDON, ROBERT, Chancellor of Moray.

Quarterly, as before ; foliage at the top and sides of the shield.

" S' M. ROBERTI GORDON CHANCELLARII DE MO," A.D. 1580. —*Duff Charters.*

432. GORLEY, ALAN.

Device of a squirrel in centre of two squares interlaced.

" S' ALANI GORLEY."—*Appended to the same Homage as No. 48, A.D. 1296.— H.M. Record Office.*

433. GORTHIE, TRISTAM, of that Ilk.

A chevron between three fleurs-de-lis ; foliage around the shield.

" S' TRISTAEM GORTHIE "—*Appended to Resignation of some lands of Balmakin A.D. 1536.—Crawford and Balcarras Charters.*

434. GORTHIE, GEORGE, of " Gortus," Son of the above Tristam.

A chevron between three fleurs-de-lis.

" S' GEORGE GORTHS OF YAT ILK." —*Appended to the same Instrument as the last.*

435. GOVERT, EDWARD.

An escallop shell, not on a shield.

" S' EDWARDI DE GOVERT." Detached Seal.—*H.M. Record Office.*

436. GRAHAM, ALEXANDER.

A chevron ; in base an escallop shell, and on a chief two of the same.

" S' ALEXANDRI GREAM." A.D. 1537.—*The late George Smythe, Esq.*

437. GRAHAM, DAVID, " Dni de Dundaff."

On a chief, three escallop shells ; the shield in centre of elegant tracery

" SIGILLUM DAVID DE GRAME."—*Appended to Appointment of Plenipotentiaries, 26th September 1357.—H.M. Record Office.*

438. GRAHAM, PATRICK, Son and Heir of Sir David Graham of Dundaff.

Much injured. On a chief, three escallop shells, with a label apparently ; the shield surrounded by tracery.

K

s' PATRICII GRAME."—*Appended to an Indenture for the Maintenance of a Chaplain at the Holyrood Altar in the Church of Dumbarton, 10th February 1372.—Records of the Burgh of Dumbarton.*

439. GRAHAM, HENRY.

An antique gem.   Victory in a car.
" s' HENRICI DE GRAEM."   Detached Seal.—*H.M. Record Office.*

440. GRAHAM, JAMES, MARQUIS OF MONTROSE.

A device, not heraldic, though on a shield; two rocks, with a gulf between; on the top of one is a lion about to spring forward, emblematic of the uncompromising principle of Montrose's own loyalty and advice to King Charles II.

For a full and interesting notice of this device see Mr. Napier's able work, " *Memoirs of Montrose,*" A.D. 1854. vol. ii. p. 746.

Motto, " NIL MEDIUM."—*From a Letter of the Marquis to the Earl of Seaforth. 26th March 1650.—Mark Napier, Esq.*

The accompanying illustration is twice the size of the original seal.

441. GRAHAM, JAMES, THIRD MARQUIS OF MONTROSE.

A fine seal, but now much injured.   Quarterly: first and fourth, on a chief three escallop shells for Graham; second and third, three roses for the title of Montrose; above the shield a coronet of five points.   Crest, on a coronet above a full-faced helmet with mantlings, an eagle preying on a stork.   Supporters, two storks.

" SIGILLUM JACOBI GRAM MARCHIONIS MONTIS ROSANI."—*Appended to Charter by the Marquis to Sir Colin Campbell of Aberuchil of lands of Dalvinoth, etc., 26th February 1675.—David Laing, Esq.*

442. GRAHAM, JOHN, EARL OF MENTEITH.

Quarterly: first and fourth, three escallop shells for Graham; second and third, per fess, in the first two chevrons, for Stratherne; the second chequé no doubt meant for the Stuart fess; slight foliage surrounds the shield.

" s' JOHANNIS GRAHEM COMITIS DE MENTEITH."—*Appended to Sasine of Drumlaw, etc., in Perthshire, in favour of Mariota Campbell, daughter of John Campbell of Glenurquhay, 2d November 1551.—Breadalbane Charters.*

443. GRAHAM, MARGARET.

A fess chequé, and in chief three escallop shells.

"s' MARGARET GRAHAM."—*Appended to Perambulation of Brockholes, A.D. 1431. - Dean and Chapter of Durham.*

George "Gram" is one of the perambulators, perhaps a brother of this Margaret whose seal he has used for the occasion.

444. GRAHAM, PATRICK.

On a chief three escallop shells; at each side of the shield is a boar's head.— *Appended to General Release given by John Balliol to King Edward I., 2d January 1292.—H.M. Record Office.*

445. GRAHAM, ROBERT, of KINROST.

Couché: on a chief engrailed three escallop shells. Crest on helmet, a stag's head cabossed. Supporters, two lions rampant guardant.

"s' ROBERTI GRAM."—*Appended to Sasine of the lands of Keir to Lucas Stirling, by Norman Lesley of Rothes, A.D. 1433.—Keir Charters.*

446. GRAHAM, ROBERT, of WACHINGTON.

An escallop shell, in chief two roses (?).

"SIGILL. ROBERTI GRAHAM DE WACHINGTOUN," (?) A.D. 1620.—*Kilsyth Charters.*

447. GRAHAM, ROBERT, of KNOCKDALLIAN.

Very rude work. Two escallop shells in chief, and in base a cross-crosslet fitchée issuing from a crescent; foliage at the top and sides of the shield.

"s' ROBERTI GRAHAM DE KNOCKDOLIAL." Detached Seal.—*W. Fraser, Esq.*

448. GRAHAM, WILLIAM.

A boar's head, not on a shield.

"s' WILL. DE GREM." Detached Seal. –*H.M. Record Office.*

449. GRANT, ALEXANDER.

On a fess, between three crowns, a star.

"s' ALEXANDER GRANT."—*Appended to Retour of Hew Fraser as son and heir of Thomas Fraser of Lovat, his father, 10th January 1524.—Lovat Charters.*

450. GRANT, JOHN, of Culcalocu.

A fess between three crowns.

"s' johannis graunt"—*Appended to Reversion by J. Grant to Hew Lord Fraser, of the lands of Ester Leis, 9th November 1529.—Lovat Charters.*

451. GRAY, ANDREW.

A lion rampant crowned. Crest on helmet and coronet, a swan's neck.

"sigillum andree gray." Very rude seal.—*Appended to Letter of Andrew Gray, consenting to his son's remaining as hostage, 28th March 1424.—H.M. Record Office.*

452. GRAY, ANDREW, Lord.

A lion rampant, within a bordure engrailed.—*Appended to Treaty between King Henry Sixth and King James Second, 15th November 1449.—H.M. Record Office.*

453. GRAY, ANDREW, Lord.

A lion rampant within a bordure engrailed.

"sigillum andre dni greye."—*Appended to Treaty of Alliance between Alexander Duke of Albany and King Edward IV. against the King of Scots (James Third), 11th February 1482.—H.M. Record Office.*

454. GRAY, PATRICK, Lord.

A lion rampant within a bordure indented.

"s' patricii dni grey." Very rude. Detached Seal.—*H.M. Record Office.*

455. GRAY, ROBERT, Master of the Mint of Scotland.

On a bend a mullet. Crest on helmet and coronet, a boar's head muzzled.

"s' roberti [gray]."—*Appended to Resignation by John Hole, of Leith, of some lands in Coldingham to the Prior of St. Ebbe, Coldingham, A.D. 1429, and having no proper seal of his own, he uses that of Robert Gray, of Leith. "Magistri monete," and Burgess of Edinburgh.—Dean and Chapter of Durham.*

456. GREIG, ROBERT.

An annulet.

"s' roberti greig."—*Appended to a Charter by Robert Greig of an annual rent of forty-four marks from the eastmost half of the "Cuningares," to Martyn Geddys and Marion Monypene, his spouse, 6th June 1582.—University of St. Andrews Charters.*

457. GUTHRIE, ALEXANDER, of Kynealdrum.

> Very indistinct. Quarterly: a garb in each quarter; or it might be blazoned, a cross cantoned with four garbs.
>
> " *s' ALEXANDRI DE GUTHRIE.*"—*Appended to same Instrument as No.* 379.

458. GUTHRIE, JANET, Wife of John Matheson.

> Quarterly: first and fourth, a lion rampant for Guthrie; second and third, a garb for Cumin (?); the shield prettily ornamented with foliage.
>
> " *s' JONET GUTHRIE.*"—*Appended to a Charter of a Mill at the Water of Leith, 25th July* 1563.—*Edinburgh Charters.*

459. GUTHRIE, JANET, Relict of Mr. James Lawson or Lowson, Minister of Edinburgh.

> Quarterly: first and fourth, three garbs for Cumin (?); second and third, a lion rampant for Guthrie; foliage at the top and sides of the shield.
>
> " *s' JONETE GUTHRIE.*"—*Appended to a Reversion by Janet Guthrie, relict of Mr. J. Lawson, minister of Edinburgh, and James Lawson, his son, with consent of Mr. John Lindsay, parson of Menmuir, to Sir David Lindsay, Knight, of an annual rent of 100 marks, deponed to her out of the lands of Meikle and Little Menzies, etc.,* A.D. 1586.—*Crawford Charters.*

460. GUTHRIE, ROBERT, of Kinbletumont.

> Very indistinct, but apparently a bull's (?) head couped between three garbs: the shield supported by an angel.
>
> " *s' ROBERTI GUTHRIE DE KINBLATUMO.*"—*Appended to Charter by Richard Guthrie to Master Peter Young, "Elemosinair" to the King, James VI.,* A.D. 1580. —*Aldbar Charters.*

461. HAGERSTON, HUGH DE, son of John.

> A rose or cinquefoil, not on a shield.
>
> " *s' HUGONIS DE HAGARSTN.*"—*Appended to Charter of Ballesdon to the Monks of St. Cuthbert's of Durham.*—*Dean and Chapter of Durham.*

462. HALDANE, ANDREW.

> A flower or wheel ornament, not on a shield.
>
> " *s' ANDREI DE HALDANIS.*" Detached Seal.—*H.M. Record Office.*

463. HALIBURTON, SIR WALTER, Lord Dirlton, created Lord Haliburton A.D. 1440.

> Quarterly: first and fourth, on a bend, three mullets for Haliburton; second, three bars for Cameron; third, a bend for De Vaux of Dirlton.—*Appended to the same Instrument as No.* 11.

464. HALIBURTON, WILLIAM.

A fess between two mascles in chief, and a human heart in base.

" s' WILLMI DE HALIBURTON."—*Appended to Transumpt of Charter by Sir David Brown, Lord of Combereolstown, to the Abbey of Holyrood,* 1466.—*Ballencreif Charters.*

465. HALL, ARTHUR, OF FULLBAR, RENFREW.

A fess chequé of two tracts between three cranes' heads erased.

" s' ARTHURI HALL DE [F']ULLBAR."—*Appended to Charter of Fullbar to William Hall, his son,* 20th August 1559.—*J. Hall Maxwell, Esq.*

466. HAMILTON, WALTER.

Three roses (cinquefoils ?) and a label.

" s' WALTERI D[E HAMI]LTON." Detached Seal.—*H.M. Record Office.*

467. HAMILTON, JAMES, LORD.

Three cinquefoils pierced.

" s' JACOBI DNI HAMILTON."—*Appended to Truce,* 28th August 1444.—*H.M. Record Office.*

468. HAMILTON, JAMES, FIRST EARL OF ARRAN.

Quarterly: first and fourth, a galley or lymphad for Arran; second and third, three cinquefoils pierced for Hamilton.

" s' JACOBI DNI HAMILTON COMITIS DE ARRANIE."—*Appended to Ratification of a Truce for one year between King Henry VIII. and King James V.,* 7th October 1518. —*H.M. Record Office.*

469. HAMILTON, JAMES, LORD, ETC. SAME PERSON AS THE LAST. *Plate* VI. *fig.* 4.

Couché. Quarterly: first and fourth, Hamilton; second and third, Arran, as before. Crest on helmet, an oak-tree with a frame-saw fixed transversely in its trunk; the background ornamented with foliage.

" s' JACOBI COMITIS DE ARRANI AC DOMINI HAMILTO."—*Appended to Charter of lands in Lanark to Gavin Hamilton of Kirkle and John his son and heir apparent,* 28th May 1525.—*Smeaton Charters.*

470. HAMILTON, ALEXANDER, FIAR OF INNERWICK.

A buckle between three cinquefoils pierced.

" s' ALEXANDER HAMILTOUN."—*Appended to Charter of the lands of Bawber, etc., in Renfrew, to Gilbert Lauder of Bawber,* 20th February 1539.—*Majoribanks Charters.*

171. HAMILTON, ALEXANDER, of INNERWICK.

A small seal, oval shaped. Quarterly: first, second, and third, a
cinquefoil with a base chequé of two tracts in the second
quarter, and a chief of the same in the third for Hamilton
and Stuart; fourth, a buckle, tongue in base, for De Glay of
Innerwick; the shield between the initials A · H · —. *Appended
to a Precept of Sasine, 3d December* 1588.—*Blackbarony
Charters.*

This marshalling of the coats of Hamilton, Stuart of Cruixton, and De Glay, differs
materially from that given by Nisbet, viz., gules: a fess chequé argent and azure
(Stuart of Cruixton) between three cinquefoils, ermine (Hamilton), all within
a bordure of the last, charged with eight buckles of the third (De Glay of
Innerwick). Here the cinquefoils (more like mullets) of Hamilton are placed
singly in separate quarters; the Stuart fess is converted into a base chequé
in the second, and a chief in the third quarter, and the buckle of De Glay
occupies the fourth,—a most extraordinary and unheraldic arrangement, only
to be explained by supposing the engraver has taken unwarrantable liberties
with the charges in order to adapt them to the limited space.

172. HAMILTON, SIR THOMAS, CREATED EARL OF MELROSE, 1619; EXCHANGED THAT
TITLE FOR EARL OF HADDINGTON, 1627. *Plate* VII. *fig.* 7.

A fine seal, though slightly damaged. Quarterly: first and fourth, a fess, wavy,
between three roses, for title of Melrose; second and third, on a chevron
between three cinquefoils, a buckle, tongue erect, within a bordure charged
with eight thistles, leaved and slipped for Hamilton of Innerwick. Crest on
helmet, with mantling, above an Earl's coronet, two dexter hands joined fesswise, issuing from clouds, holding between them a branch of laurel erect:
motto on a scroll issuing from the crest, " PRÆSTO ET PERSTO." Supporters,
two greyhounds, collared.

SIGILLUM THOME COMITIS DE HADINGTON [DOMINI BY]RIS ET BINNING."—*Appended to
a Charter by Thomas Earl of Haddington to George Brown of Colston and
James Brown his son, of the tithes of the Barony of Colston, 20th May* 1633.
—*Colston Charters.*

173. HAMILTON, THOMAS, EARL OF HADDINGTON, THE SAME PERSON AS THE LAST.

A small signet, and remarkably fine example of the art of the period. On a
chevron between three cinquefoils, a buckle, tongue erect, within a bordure

charged with eight thistles leaved and slipped; above the shield an Earl's coronet, and around it the initials, $_{T \cdot H}^{E}$ (Thomas Earl of Haddington).— *From a Letter addressed to Sir Richard Kerr of Ancrum, dated 24th December 1634.—David Laing, Esq.*

474. HAMILTON, ... of Hagg.

A salmon's head couped, with a ring through its nose, between three cinquefoils. Crest on a helmet with mantlings; an oak-tree with a frame-saw fixed transversely in the trunk; on a ribbon above, the word "through." The crest of this family, as recorded in the Lyon Register, is a salmon haurient.

From a steel seal having three faces, the first as above, the next having the crest (the oak-tree) and motto only, and the third occupied with a singular device, probably bearing some allusion to the habits or pursuits of the original owner, who was perhaps the founder of the family. The device is a man sitting; his left hand, holding a roll of paper, rests on a table, on which is at one end an inkstand and pen, and at the other a few coins or medals; surrounding the device is the motto "FIDES PRÆSTANTIOR AURO." The costume of this figure, as well as the style of ornament in which the seal is made, would point to the period of the Revolution as that in which it was fabricated and used.—*This interesting seal is in the possession of J. Buchanan Hamilton, Esq. of Lenny and Bardowie.*

475. HASWELL, WILLIAM.

A lion rampant in centre of a rose, or rather the shield is of a rose shape. A very pretty seal.

" S' WILE[L]MI DE HESWEL."   Detached Seal.—*H.M. Record Office.*

476. HATELY, ROBERT.

A bird passant, not on a shield.

" SIGILL. ROBERTI L'PORTE."—*Appended to Charter of Confirmation by Robert the Porter and Matilda his wife, of lands at Fannes and a toft at Mollerstane, to the Prior and Monks of St. Ebbe at Coldingham.—Dean and Chapter of Durham.*

477. HAUVIE (HARVEY?), ROBERT.

Device of a lion coiled in centre of two squares interlaced.

" S' ROBTI DE HAUVIE."—*Appended to the same Homage as No. 48, A.D. 1292.—H.M. Record Office.*

478. HAY, NICOLAS, second Earl of Errol.

Couché; three escutcheons. Crest, on helmet with mantlings, a stag's head.

"s' nich[olas] comit. [de errol]."—*Appended to Precept of Sasine for infefting Gilbert Hay in the lands of Cry. Aberdeen, 12th August 1467.—Communicated by W. Fraser, Esq.*

479. HAY, WILLIAM, fifth Earl of Errol.

Couché; three escutcheons. Crest. on helmet with mantlings, a hart's head couped

"s' willielmi comitis de errole."—*Appended to "A Bond given by William Earle of Errol, to John Lindesaye, Earle of Crauford, quherin the said Earle William binds and oblisses him, by the faith of his bodey, to resigne the schriffschipe of Aberdeine. in his hands, back againe at quhat tyme the said Earle Johne, or any gentilman bearing the surname of Lindesaye, repaying alwayes to the said Earle William the somme of six hundereth marks Scotts money;" dated at Edinburgh, 6th February 1509.—Hutton's "Sigilla," p. 132.*

480. HAY, WILLIAM, Earl of Errol.

Couché to sinister; three escutcheons. Crest, on helmet with mantlings, a ram's head.

"s' willelmi hay [comitis] de erroll." Detached Seal.—*H.M. Record Office.*

481. HAY, GEORGE, sixth Earl of Errol.

Couché; three escutcheons. Crest on helmet, an eagle's head; foliage in background.

"s' georgii comitis de erol."—*Appended to Precept of Sasine for infefting John Hay his son in the lands of Nether Muchals, etc., in Kincardine, 18th September 1542.—Communicated by W. Fraser, Esq.*

482. HAY, FRANCIS, ninth Earl of Errol.

Couché; three escutcheons. Crest, on coronet above a helmet with mantlings, an eagle's (or falcon's) head erased, surmounting an ox yoke. Motto, on ribbon below the shield, "serva jugum."

"s' francisci comitis errolle dni hay constabularii scotie."—*Appended to a Charter by the Earl of Errol to Francis Hay, his third son, of the lands of Tawertie, in barony of Slains, Aberdeenshire, A.D. 1625.—Communicated by George Logan, Esq., Teind Office.*

483. HAY, SIR GEORGE, Knight, Viscount Dupplin, created Earl of Kinnoull A.D. 1633, Lord High Chancellor of Scotland. Died 1634. *Plate VII. fig. 2.*

A fine seal. Shield, quarterly: first and fourth, a unicorn salient, within a bordure

L

charged with eight thistle-heads and roses dimidiated, being a coat of augmentation; second and third, three escutcheons, for Hay. Above the shield is a
coronet of nine points.

SIGILL. GEORGII VICECOM. DUPLINLE BAR. HAY DE KINFAUNS ET MAG. SCOTLE CAN
CELLAR."— *Appended to Charter of Confirmation by George Earl of Kinnoull
to William Sibbald, of three and a half acres of land in the barony of Kinnoull,
co. Perth. 24th February 1634.—In possession of G. R. Mercer, Esq. of
Gorthy.*

484. HAY, HUGH.

Three escutcheons within a bordure engrailed.
" s' HUGONIS DE HAIA." Detached Seal.—*H.M. Record Office.*

485. HAY, JAMES.

Three escutcheons. A cinquefoil in centre.
"SIGILLUM JACOBI HAYA." Detached Seal.—*H.M. Record Office.*

486. HAY, JOHN, of LOCHLOY.

Rude and imperfect. A lion's head (?) erased between three escutcheons
" s' JOHANNIS HAY DE LOCLOY."—*Appended to the same Instrument as No.* 271

487. HAY, MARIOTE, WIFE OF ALEXANDER LAUDER.

Three escutcheons.
" s' MARIOTE HAY."—*Appended to the same Instrument as No.* 90.

488. HAY, THOMAS.

An escutcheon in chief, and a buckle, the tongue fesswise, in base.
" s' THOME HAY."—*Appended to Precept of Sasine of the lands of Meikle Geddes, etc.,
in the Earldom of Moray, sheriffdom of Nairn,* 1493.—*Cawder Charters.*

489. HAY, WILLIAM.

Three escutcheons. Crest on helmet, a stag's head. Supporters, two lions
" s' WILLELMI HAYA."—*Appended to Letter of William Hay, consenting to his son's
remaining as hostage,* 31st March 1424.—*H.M. Record Office.*

490. HAY, WILLIAM, of NAUGHTON.

Couché; three escutcheons within a bordure, indented. Crest, on helmet, a mer-

maid holding a mirror in her right hand ; the background ornamented with trees
and a stream of water.

"s' willia hay."—*Appended to an Instrument dated* 1467.—*Communicated by
W. Fraser, Esq.*

491. HEPBURN, PATRICK.

Very fine work. Conché. On a chevron, a rose between two lions rampant respecting. Crest, on helmet, a camel's head bridled and belled (?). Supporters, two
lions rampant. Background foliated.

"s' patricii hepburn dni de halis."—*Appended to Charter of four acres of land
in Coldingham to the Monks of St. Ebbe at Coldingham,* 24th *November* 1450.
—*Dean and Chapter of Durham.*

492. HEPBURN, PATRICK, created Earl of Bothwell 5th October 1488, and
Lord High Admiral of Scotland.

A small seal. A device, not on a shield, of a three-legged stool, and above it the
word "keiy," or probably a monogram of "keep tryste"—the family motto,
as in No. 494.—*Appended to an Indenture for Peace between King Henry VII.
of England and King James IV. of Scotland,* 24th *January* 1501.—*H.M.
Record Office.*

493. HEPBURN, PATRICK, third Earl of Bothwell. *Plate III. fig. 7.*

Conché. On a chevron, a rose between two lions rampant respecting ; in base
an anchor. Crest, on helmet with mantlings, a horse's head.

"s' patrich hepburn, admiral sco."—*Appended to Presentation of Archibald Eilem
to the Prebendary of Auldhamstocks,* 27th *May* 1515.—*Home Charters.*

494. HEPBURN, JAMES, fifth Earl of Bothwell, Husband
of Queen Mary.

A shield bearing an anchor and the family motto, "keep
tryste;" the latter word as a monogram. Around
the shield are the initials $_{I. B.}^{E}$. (James Earl Bothwell).

This is the official seal of Earl Bothwell as Great Admiral
of Scotland, and is impressed on paper, being a Commission by James Earl
Bothwell, Sheriff of Berwick, appointing deputies to pursue and apprehend
certain persons named therein as rebels against her Majesty. Signed by the
Earl, and dated at Craigmillar, 29th November 1566.—*In possession of Sir
Archibald Edmonstone of Duntreath, Baronet.*

495. HEPBURN, LADY JANET, Daughter of Patrick, first Earl of Bothwell and Widow of George, fifth Lord Seton.

Couché; per pale, dexter Seton; sinister Hepburn. Crest, on helmet with mantlings, a female head affronté.

"SIGILLUM JANETE HEPEN ET MARIT."—"*Procuratory given by Domina Janeta de Hepburne Domina de Settone vidua et relicta Georgii quondam Domine Settone, for the uplifting of some annual-rents deir to her, etc., of the dait 20 Apprillis 1541. This scall is appencit.*"—*Hutton's " Sigilla," p. 171.*

496. HEPBURN, JOHN.

On a chevron, surmounted by a bend, a rose between two lions rampant respecting.

"s' JOHANNIS HEPBURN."—*Appended to Charter by Thomas Fyfe, 19th August 1521. — Charters of St. Mary's College, St. Andrews.*

497. HEPBURN, JOHN, of Rollandston.

On a chevron, a rose between two lions rampant respecting; in the dexter chief point a mullet.

"SIGILLUM JOANNES HEPBURN," A.D. 1507.—*Elibank Charters.*

498. HEPBURN, JOHN, of Beniston.

Hepburn, as before, with a mullet in base for difference.

"s' JOHANIS HEPBURN DE . . ."—*Appended to Retour of George Broun of Colston, as heir to his father, George Broun, in the lands and barony of Colston, in shire of Haddington, 20th January 1558.—Colston Charters.*

499. HEPBURN, PATRICK, of Boltoun.

On a chevron, a rose between two lions rampant respecting; in base an anchor; above the shield a helmet, but no crest.

"s' PATRICII HEPBURNE."—*Appended to Precept for Sasine by P. Hepburn and his spouse, Alison Home, of the lands of Plewland, in the village of Lintoun, barony of Halis, co. Haddington, in favour of William Cockburn, de eodem, 23d July 1515.* Hepburn was formerly Master of Hailes and presumptive heir to the Earl of Bothwell.—*David Laing, Esq.*

500. HEPBURN, PATRICK, of Watchton.

Couché. Quarterly: first and fourth, Hepburn; second, a fess engrailed for Sinclair;

third, au orle for Landel; and on a chief, three mullets for Rutherford. Crest on helmet with mantlings, an antelope's head.

" s' PATRICII HEPBURN DE VACHTON." —*From the Collection of General Hutton.*

501. HETLYN, JOHN.

A wheel ornament, or flower of nine leaves.

"s' JOHS DE HETLYN."—*Appended to same Homage as No. 48, A.D. 1292.—H.M. Record Office.*

502. HINFORT, ISAAC.

Device of a greyhound seizing a hare.

" s' YSAAC DE HINFORT."—*Appended to the same Instrument as the last.*

503. HINTOUR, or KINTORE, WALTER.

Device of a pair of scissors open.

" s' WALT. DE KINTORE" (?).—*Appended to the same Instrument as the last.*

504. HOME, DAVID.

A lion rampant. Coarse work. Foliage at the top and sides of the shield.

" s' DAVID DE HOME."—*Appended to Letter of W. Drax, Prior of Coldingham, confirming Indenture, 1424.—Dean and Chapter of Durham.*

505. HOME, DAVID.

A lion rampant.

" s' DAVID DE HOME."—*Appended to a Retour of Simon Ward of Raynton, A.D. 1426. —Dean and Chapter of Durham.*

506. HOME, SIR DAVID, KNIGHT.

Couché; a lion rampant. Crest on helmet, a fawn's head. Supporters, two parrots.

" SIGILLUM DAVID HOME."—*Appended to an Inquisition of Patrick Blackburn, 15th October 1430.—Dean and Chapter of Durham.*

507. HOME, ALEXANDER.

Quarterly: first and fourth, a lion rampant for Home; second and third, three papingoes for Pepdie. Crest on helmet, a fawn's head; background diapered, of a lozenge pattern. Very pretty.

" s' ALEXANDRE DE HOME."—*Appended to Inquisition of the lands of John Harker, 1437.—Dean and Chapter of Durham.*

508. HOME, SIR ALEXANDER, Knight.

    Couché. Quarterly, as in the last. Crest, on helmet, head of a fallow-deer. Supporters, two lions.

    " s' ALEXANDRI DE HUME."—*Appended to Charter to the Collegiate Church of the Virgin at Dunglass, of the lands of Chirnside, co. Berwick, 5th August 1450.—Home Charters.*

509. HOME, SIR ALEXANDER, Knight, Son of the above Sir Alexander Home, created Lord Home, 1473.

    Couché. Quarterly: Home and Pepdie as in the last; crest and supporters the same. Very rudely executed.

    " s' DNI ALEXANDRI HUME."—*Appended to Charter to Thomas Home of Crowdy, 30th July 1486.—Home Charters.*

510. HOME, ALEXANDER, Grandson of the above Lord Home, and afterwards second Lord Home.

    Quarterly, as before, with a mullet for difference. No crest or supporters.

    " s' ALEXANDRI HOUM."—*Appended to the same Instrument as the last.*

511. HOME, ALEXANDER. Same person as the last.

    Quarterly, as before; over all an escutcheon bearing an orle for Landel.

    " s' ALEXANDRI HUME."—*Appended to Charter of the lands of Ninewells to Patrick Hume of Fastcastle, 9th September 1485.—Home Charters.*

512. HOME, GEORGE, fourth Lord, Grandson of the above Alexander. *Plate* VIII. *fig.* 6.

    Couché. Quarterly, as in the last. Crest, on helmet with mantlings, a fallow-deer couchant.

    " s' GEORGI DOMINI DE HOUME."—*Appended to Precept of Sasine by Alexander Lord Home, " with consent of his father, George Lord Home," in favour of Lady Elizabeth Hamilton, of the Mills of Broxmyline. A.D. 1542.—Home Charters.*

513. HOME, ALEXANDER, fifth Lord, Son of the above George. *Plate* VIII. *fig.* 5.

    Couché. Quarterly: first, Home; second, Pepdie; third, a unicorn rampant for Samuelston; fourth, on a bend three mascles for Haliburton; over all, on an escutcheon, an orle for Landel. Crest, on helmet with mantlings, a fallow-deer.

    " s' ALEXANDRI DOMINI DE HOUME."—*Appended to the same Instrument as the last.*

514. HOME, ALEXANDER, sixth Lord, Son of the above Alexander. Created Earl A.D. 1604.

Couché. Quarterly; Home and Pepdie, with Landel on an escutcheon. Crest, on helmet with mantlings, a lion's head erased. Supporters, two lions rampant.

" sigillum alexandri domini home."—Appended to Charter of the lands of West Fenton to Alexander Home of North Berwick, A.D. 1591. —Home Charters.

515. HOME, DAVID, of Wedderburn.

Quarterly: first and fourth, a cross engrailed for Sinclair of Polwarth; second, a lion rampant for Home; third, three papingoes for Pepdie.

" s' david houm d' wodbrn," A.D. 1514.—Kilsyth Charters.

516. HOME, SIR GEORGE, of Wedderburn, Knight.

Quarterly: first and fourth, a lion rampant for Home; second, three papingoes for Pepdie; third, a cross engrailed for Sinclair of Polwarth. A rose and thistle ornament at the top and sides of the shield.

" s' d. georgii home de voderbu,"—Appended to Charter of Sir George Home to Margaret Dunlop, in Eyemouth, 21st July 1611.—H.M. General Register House.

517. HOME, DAVID, of Wedderburn. Son of the above Sir George Home.

The same charges as in the last, but the quarters are transposed, the Home quarter being here the second and third, Pepdie first, and Sinclair fourth; foliage at the top and sides of the shield same as in the last.

" s' davidis home de [voderburn]."—Appended to Charter by Sir David Home, A.D. 1617.—H.M. General Register House.

518. HOPE, SIR THOMAS, Bart., of Craighall, Lord Advocate of Scotland.

A signet. A chevron between three bezants; the shield surrounded with scroll ornament very prettily executed, and above it the initials, $_{T \cdot H}^{s}$. (Sir Thomas Hope).—From a Letter to the Earl of Ancrum, dated 29th December 1631.—Marquis of Lothian.

519. HUNTINGTON, DAVID, EARL OF. Son of Prince Henry, and Grandson of King David I. Plate IV. fig. 2.

A much injured seal. An armed knight on horseback, with drawn sword in his right hand and a shield on his left. One or two letters only of the inscription now remain.—Appended to a Document in the Harleian Collection, British Museum. circa 1160.

HUNTLY. EARL of. Vide Gordon.

520. HUNTLY, ROBERT.

An eight-leaved flower or wheel ornament, not on a shield.

" s' ROBERTI DE HUNTL."—*Appended to the same Homage as No. 48*, A.D. 1292. — *H.M. Record Office.*

521. HUTTON, ROBERT.

A lion rampant, not on a shield.

" s' ROBERTI DE HUTTAUN."   Detached Seal.—*H.M. Record Office.*

522. HYNDMAN, HECTOR.

A saltire ; in the honour point a mullet.

" s' [HECTOR] HYNDMAN," A.D. 1578.—*Cocknirnie Charters.*

523. INGLIS, MARGARET.

Device of a pelican and young.

" s' MARGARETE ENGGLIS."—*Appended to the same Homage as No. 48*, A.D. 1292 —*H.M. Record Office.*

524. INGLIS, MARGARET.

A lion rampant ; in chief three mullets ; foliage at the top and sides of the shield.

" s' MARGARETE INGLIS."—*Appended to Procuratory for Resignation of the half of Balmakin, 2d July* 1594.—*Crawford Charters.*

525. INIRPEFFER, DAVID.

A lion coiled, not on a shield.

" s' DAVIT DE INIRPEFE."   Detached Seal.—*H.M. Record Office.*

526. INNES, ALEXANDER, OF INNES.

Couché.   Quarterly : first and fourth, three mullets ; second and third, three boars' heads and necks erased, for Aberkerdour ; a crescent as a difference.   Crest on helmet, with mantlings, a plume of feathers (?).

" s' ALEXAND. INNES."—*Appended to Contract of Marriage between Alexander Innes of that ilk, and Christian Dunbar, daughter of Sir James Dunbar of Cumnok, Knight, 4th December* 1493.—*Communicated by Cosmo Innes, Esq.*

527. INNES, ALEXANDER, of Innes.

Couché; three boars' heads couped; in
chief three mullets. Above the shield
a helmet and mantlings, but no crest.

" S' ALEXANDR INNES DE LODEM." — Appended to Precept of Sasine by Alexander Innes, son and heir of Alexander Innes of that ilk, to his cousin,
James Innes of Rothmakenzie, of lands
in Elgin and Forres, at Edinburgh, 23d
July 1542. — Duffhouse Charters (?).

528. INNES, JAMES.

Quarterly: first and fourth, three mullets
for Innes; second and third, three boars
heads couped for Aberkerdour. Foliage at the top and sides of the shield.

" S' JACOBI INNES." Detached Seal. — Cosmo Innes, Esq.

529. INNES, JAMES, of Innes.

Couché. Quarterly: Innes and Aberkerdour, as in the
last. Above the shield a helmet, but no crest;
foliage in the background.

" S' JAMES OF INNES." — Appended to a Precept by
James Innes in favour of Margaret Culane, his
spouse, 7th May 1489. — Floors Charters.

530. INNES, ROBERT, of Rothmakenzie, second Son of
THE ABOVE JAMES INNES.

Two mullets in chief, and a boar's head erased contourné in base.

" S' ROBERTI INNES." — Appended to Retour of Patrick
Hay of Ury of the annual of Kilmalumak, 6th
October 1531. — Floors Charters.

The dispensing with the Innes and Aberkerdour quarters
here seems to carry us back to the period of com-
posed coats, where, as in this, a portion of each coat
of the families united was borne on one shield. In
the present instance the arrangement has probably
been adopted to mark the cadency of the owner.

M

531. INNES, ROBERT, of Innes.

>Quarterly: Innes and Aberkerdour.

>"s' ROBERTI INNES DE EODEM."—*Appended to a Contract between Robert Innes of Innes and Alexander, Master of Elphinstone, 1592.—Floors Charters.*

532. INNES. ROBERT, of Innermarkie.

>Couché; three mullets. Crest, on helmet with mantlings, an eagle's (?) head.

>"s' ROBERTI INNES DE INNERMERKI."—*From the Collection of General Hutton.*

533. INNES, WALTER.

>Three mullets.

>"s' WALTERI INNES."—*Appended to Inquisition of the lands of Kilravock, 11th February 1431.—Kilravock Charters.*

534. INNES, WILLIAM.

>A star, not on a shield

>"s' WILLI DE YNAUS."—*Appended to Homage Deed, 10th July 1295.—H.M. Record Office.*

535. IRVINE, ALEXANDER, of Belteis.

>Two cross-crosslets fitchée, surmounted by a fess between three bunches of holly leaves.

>"s' ALEXANDER IRVIN."—*Appended to Precept of Sasine by W. Abbot, of Arbroath for infefting Alexander Irvine of Drum in the lands of Forglen, Aberdeenshire, 12th February 1483.—Communicated by Lord Lindsay.*

536. ISLES, ALEXANDER of the.

A galley, with two figures in it.

"s' ALEXANDRI DE ISLE."—*Appended to a Letter dated 7th July* 1292.—*H.M. Record Office.*

537. ISLES, ALEXANDER of the, Lord of the Isles and Earl of Ross.

Quarterly: first and fourth, a galley under sail; second and third, three lions rampant for Ross; the shield in front of an eagle. The inscription on this pretty seal is rather imperfect, but apparently is "s' ALEX-ANDRI COMITIS ROSSIE DOMINI INSULARUM."—*Appended to Charter, not dated, confirming a Charter by William Earl of Ross, A.D.* 1358.—*Duke of Richmond.*

538. ISLES, ALEXANDER of the.

Device of a lion couchant gardant, not on a shield; over the back the letters M. B.

"SECRETE · · ·"—*Appended to Letter on the Ransom of David II., 4th February* 1357.—*H.M. Record Office.*

539. ISLES, DAME MARGARET of the, Lady of the Isles and Ross.

This seal is unfortunately much injured, but has been evidently a very fine design. A figure of a lady standing beneath a Gothic canopy, supporting with each hand a shield, resting on the back of a dragon standing within the battlements of a tower, the dexter shield charged with a galley surmounted by an eagle displayed, while the sinister appears to bear three lions rampant for Ross. The inscription is quite illegible, A.D. 1420.—*Cosmo Innes, Esq.*

540. ISLES, JOHN of the, Lord of the Isles and Ross.

The same design and charges as the seal of Alexander of the Isles, his grandfather, No. 537, but rather larger, and at the sides of the shield are oak leaves.

"s' JOHIS DE YLLE COMIS ROSSIE DNI INSULARU."—*Appended to Precept of Sasine in favour of Alexander Sutherland to the lands of Ester Kynles, sheriffdom of Inverness, A.D.* 1449.—*Cawdor Charters.*

541. JAFFRAY, ALEXANDER, Bailie of Aberdeen.

Paly of six; on a fess three mullets.

"s' ALEXANDRI JAFRAI," about A.D. 1613. Detached Seal.—*The late P. Chalmers. Esq. of Aldbar.*

542. JOHN, Son of John.
    A stag's head to sinister, not on a shield.
    " s' JOANIS FIL. JOANIS."—*Appended to the same Homage as No. 48, A.D. 1292.*
    —*H.M. Record Office.*

543. JOHNSTON, SIR ARCHIBALD, Knight, Lord Warriston, Lord of Session, A.D. 1641, executed 1664.
    A small signet. A saltire, and on a chief three cushions. The crest, on helmet with mantlings, and motto surrounding, are so much damaged that an accurate description cannot be given.—*From a Letter addressed to the Marquis of Argyle, dated about 1647-8.—Marquis of Lothian.*

544. JOHNSTON, CATHERINE, Wife of John Adamson.
    A saltire cantoned, with two mullets in fess and as many crescents in chief and base, on a chief three cushions.
    " s' KATRINE JOHNSTOUN."—*Appended to an Instrument dated 24th January 1567.—Edinburgh Charters.*

545. JOHNSTON, MARGARET.
    A saltire; in upper canton a pellet, and on a chief three cushions.
    " s' MARGARETE JOHNSTOUN," A.D. 1584.—*Communicated by C. Baxter, Esq.*

546. JOHNSTON, ROBERT, Bailie of Aberdeen.
    A bend between a boar's head and neck erased in chief, and three cross-crosslets, fitchée, in base. On a chief, three cushions.
    " s' ROBERTI JOHNSTOUNE," A.D. 1617.—*Aldbar Charters.*

547. KAYCEYLGECHO, (?) GILMORE.
    A stag's head cabossed, not on a shield.
    " s' GILMORE KAYCEYLGECHO." (?,—*Appended to the same Homage as No. 48, A.D. 1292.—H.M. Record Office.*

548. KEITH, ADAM.
    A tree and a rabbit or hare, not on a shield.
    " ADE DE KEITH."—*Appended to the same Instrument as the last.*

549. KEITH, ALEXANDER.

A very pretty design, not on a shield. A lion coiled, in centre of tracery.

"s' ALEXANDRI DE KETH." Detached Seal.—*H.M. Record Office.*

550. KEITH, JOHN, SECOND SON OF SIR EDWARD KEITH, GREAT MARSHAL OF SCOT

LAND. *Plate VI. fig. 4.*

A chief paly of six; over all a bend. The shield surrounded by rich tracery.

"SIGILLUM JOANNIS DE KETH."—*Appended to Charter, by John Keith and Mariota his spouse, to the Order of the Carmelites, A.D. 1390.—Marischal College, Aberdeen.*

551. KEITH, MARIOTA, WIFE OF THE ABOVE JOHN KEITH, AND DAUGHTER AND HEIRESS

OF REGINALD CHEYNE OF INVERUGIE. *Plate VII. fig. 5.*

Keith, as in the last, impaling Cheyne, viz., on a bend, between six cross-crosslets, fitchée, three mullets. The shield in centre of elegant painted tracery.

"s' MARIOTE DE KETH."—*Appended to the same Instrument as the last.*

552. KEITH, SIR ROBERT, KNIGHT, SECOND SON OF SIR

WILLIAM KEITH, GREAT MARSHAL.

Couché; a chief paly of six (erroneously engraved five on the seal), with a label of three points. Crest, on helmet with mantlings, a stag's head erased.

"SIGILLUM ROBERTI DE KETHE."—*Appended to Indenture between Sir William Keith and Sir Robert Keith, A.D. 1442.—Communicated by W. Fraser, Esq.*

553. KEITH, WILLIAM.

On a chief, three palets. Crest, on helmet, a stag's head.

"s' WILLELMI D' KETH." Detached Seal.—*H.M. Record Office.*

554. KEITH, WILLIAM.

A chief paly of six. Crest, on helmet, a stag's head. The shield in centre of tracery.

"s' WILLELMI DE KETH."—*Appended to Appointment of Plenipotentiaries, 26th September 1357.—H.M. Record Office.*

555. KELOR, RANDULPH.

A cross-crosslet, fitchée, not on a shield.

"*s' ranulphi d' kelor.*"—*Appended to the same Homage as No. 48, a.d. 1292.—H.M. Record Office.*

556. KENBUCKE, ARCHIBALD.

A chevron between three stag's heads cabossed.

"*s' archibaldi de kenbucke.*"—*Appended to Charter of the lands of Classmigall, in Strathern, co. Perth, to William Stirling, a.d. 1455.—Keir Charters.*

557. KENLOCHY, JOHN.

An eagle displayed, charged on the breast with a lozenge.

"*s' johannis de kenlochy.*"—*Appended to a Charter by John Kenlochy, of an annual-rent from a tenement in St. Andrews to John Carmichael, Constable of St. Andrews, a.d. 1438.—University of St. Andrews Charters.*

558. KENNEDY, GILBERT, of Cassillis.

A chevron between three cross-crosslets, fitchée, all within a double tressure fleury counterfleury. Crest, on helmet and mantlings, a stag's head.

"*s' gilberti dni kenedy de cassillis.*" Detached Seal.—*H.M. Record Office.*

559. KENNEDY, GILBERT, of Kirkmichael.

A chevron between two cross-crosslets, fitchée, in chief, and a boar's head and neck erased in base (?).

"*s' gilbert de kenede.*"—*Appended to Charter by Gilbert Kennedy to William, Carmichael, of Meadowflat, of an annual-rent from a tenement in St. Andrews, 6th May 1462.—St. Salvator's College Charters.*

560. KENNEDY, JANET, Lady of Bothwell (?).

A chevron between three cross-crosslets, fitchée; foliage at the top and sides of the shield.

"*s' jonette kenedi.*"—*Appended to a Charter by Janet Kennedy, dated 25th February 1518.—Edinburgh Charters.*

561. KER, ANDREW, of CESSFORD.

Three lozenges or mascles; in middle chief a bird (martlet?); inscription not legible.—*Appended to an Indenture for a Truce between England and Scotland, 30th January* 1520.—*H.M. Record Office.*

562. KER, ELIZABETH, LADY OF BROUGHTON, DAUGHTER OF SIR W. KER OF CESSFORD.

On a chevron three mullets; in base a griffin's (unicorn's?) head erased; at the top and sides of the shield a rose.

" S' [ELIZABETH]E KER DNE DE BROCHTON."—*Appended to Precept of Sasine by Dame Elizabeth Ker, relict of James Ballenden, Baron of Broughton, to her son, William Ballenden of Broughton (created Lord Ballenden in* 1661), A.D. 1621.—*H.M. General Register House.*

563. KER, HENRY.

Oval shape, a bird standing, not on a shield.

" S' HENRICI KER." Detached Seal.—*H.M. Record Office.*

564. KER, MARY, LADY MELROSE.

On a chevron, three mullets; in base a unicorn's head contourné.

[" S' MARIE] KER."—*Appended to Charter by James Douglas, Commendator of Melrose, and Mary Ker, his spouse, to the Feuars of Gattonside and Westhouse, 28th February* 1590.—*John Alexander Smith, M.D., Sec. Scot. Antiq. Soc.*

565. KER, MERION.

On a chevron, three mullets; in base a unicorn's head erased contourné.

" MERION KAR."—*Appended to the same Instrument as No.* 27.

566. KER, SIR ROBERT, OF ANCRUM, CREATED EARL OF ANCRUM 24TH 1633.

A very pretty signet. On a chevron, three mullets. Crest, on helmet with mantlings, a stag's head erased.—*From a Letter dated 6th August* 1624.—*Marquis of Lothian.*

567. KER, ROBERT, THE SAME PERSON AS THE LAST.

A signet. Shield, quarterly : first and fourth, on a chevron three mullets for Ker; second and third, ermine; on a chief, a lion passant gardant for title of Ancrum.

A palm branch at each side of the shield, and an earl's coronet above it.—*From a Letter to his son, the Earl of Lothian, 9th December* 1651.—*Marquis of Lothian.* There is another seal of this Earl to a letter of a later date, but it differs in nothing from the above except being a little larger.

568. KER, WILLIAM, Son of Robert Earl of Ancrum, married Anne, Heiress of the Earl of Lothian, created Earl of Lothian and Newbattle, a.d. 1631.

A signet. Quarterly : first and fourth, the sun in its splendour, for title of Lothian; second and third, on a chevron, between three mascles in chief, and a unicorn's head erased in base, for Ker ; above the shield an earl's coronet, and at the sides a branch of palm.—*From a Letter to his father, the Earl of Ancrum, dated 7th March* 1631.—*Marquis of Lothian.* There are two other seals of this Earl of a later date, but the charges and design are exactly the same, only being rather of a larger size.

569. KER, ROBERT, Second Earl of Lothian.

A fine large seal, but now much broken. Quarterly : first and fourth, the sun in splendour, for title of Lothian ; second and third, on a chevron, three mullets ; in base a unicorn's head erased, for Ker. The crest, which is lost, has been on a helmet with mantlings. Supporters, two angels crowned, with a nimbus, wings expanded ; all on a ground or compartment, the only instance of it in this family.—*Appended to Procuratory for Resignation by John Murray of Blackbarony to Robert Earl of Lothian,* 1617.—*Blackbarony Charters.*

570. KER, ROBERT, Earl of Roxburgh, etc., created Earl of Roxburgh, Lord Ker of Cessford, Caverston, etc., a.d. 1616.

A fine example of the art of the period. Quarterly : first and fourth, on a chevron between three unicorns' heads erased, as many mullets, for Ker of Cessford ; second and third, three mascles for Vipont ; above the shield an earl's coronet. Crest, on three-quarter-faced helmet with mantling, a unicorn's head erased. Motto, on a ribbon issuing from the wreath, "PRO CHRISTO ET PATRIA DULCE PERICULUM." Supporters, two savages wreathed about the waist with laurel, each holding a club in the exterior hand ; the dexter resting it on the shoulder, the sinister on the ground.

"S' ROBERTI COMITIS ROXBROUGH DOMINI KER DE CESFURD ET CAVERTOUN."—*Appended to Charter of Langeroft, near Linlithgow, to Thomas Edward, merchant-burgess of said burgh, and Margaret Bell his spouse, 20th November* 1655.—*Colston Charters.*

571. KER, THOMAS.

A small signet. On a chevron, three mullets; in base a unicorn's head couped. Crest, on helmet with mantlings, a hand holding a sword sinister bendways.— *From a Letter to the Earl of Roxburgh, dated 5th May 1635.—Marquis of Lothian.*

572. KINCAID, EDWARD, Sheriff-Depute of Edinburgh.

A fess ermine, between two mullets in chief and a castle triple-towered in base.

" s' EDWARTIUS KINCAID."—*Appended to Sasine of lands at Ratho in favour of William Hatton, A.D. 1521.—Hutton Collection.*

573. KINGHORN, WILLIAM.

An eagle with wings expanded, not on a shield.

" s' WILL. DE KINGARN CLRICO." Detached Seal.—*H.M. Record Office.*

574. KINLOCH, DAVID.

A chevron between two mascles (lozenges?) in chief, and a boar's head and neck erased contourné; in middle chief, a mullet for difference.

" s' DAVID DE [KIN]LOCH."—*Appended to Charter by David Kinloch, son of " dni Henrici de Kynlouch miles," to John Carmichael, constable of St. Andrews, of a tenement in St. Andrews, A.D. 1418.—St. Salvator's College Charters.*

575. KINLOCH, DAVID, one of the Ministers of Edinburgh.

A bull's (?) head couped between two lozenges in chief, and a mullet in base.

" s' DAVID KINL[OCH]."—*Appended to the same Instrument as No. 195.*

576. KINLOCH, HENRY, Son of the above David Kinloch.

A boar's head erased, between two mascles in chief, and a mullet in base.

" s' HENREC KINLOCH."—*Appended to the same Instrument as the last.*

577. KINNAIRD, RALPH.

A saltire cantoned, with four cross-crosslets, fitchée.

" s' RADULFI DE KINARD."—*Appended to Homage Deed, 1295.—H.M. Record Office.*

578. KINNAIRD, REGINALD, of Inchture.

A saltire cantoned with four crescents.

" SIGILLUM REGN. DE KYNARDE," A.D. 1125.—*The late George Smythe, Esq.*

N

579. KINNAIRD, RICHARD.

An indistinct figure of a bird (?).

" SIGILL. RICARDI K[YN]NART."—*Appended to Charter by Richard Kynnart to John, son of Richard of Invirtuyl, of the lands of Dunort, with his sister Isabel, in free marriage. Twelfth century.—Advocates' Library.*

580. KINROSS, JOHN.

Shield-shaped seal. A lion rampant.

" S' JOHANNIS DE KINROS." Detached Seal.—*H.M. Record Office.*

581. KINTORE, JOHN.

Oval shape. A wheel ornament.

" S' JOH. DE KENONTOIR" (?).—*Appended to the same Homage as No. 48, A.D. 1262.—H.M. Record Office.*

582. KINTORE, JOHN.

Couché; a chevron between three castles or towers embattled, and on a chief three mullets. Crest on helmet, an eagle devouring its prey.

" S' JOHANNIS DE KINTOR."—*Appended to the same Instrument as No. 197.*

583. KIRK, ALEXANDER, ONE OF THE BAILIES OF ST. ANDREWS.

A saltire couped at the upper part; in base a cinquefoil.

" S' ALEXANDRI KIRK."—*Appended to an Instrument of Sasine, dated A.D. 1520.—St. Leonard's College Charters.*

584. KIRKALDY, ELIZABETH.

Two crescents in chief, and a mullet in base.

" S' ELIZA[BETH] KIRKALDI," A.D. 1582.—*Cockairney Charters.*

585. KIRKALDY, JAMES.

Two crescents in chief, and a mullet in base.

" S' JACOBI DE KIRKALDI."—*Hutton's "Sigilla," p. 169. "Sigillum Jacobi de Kirkaldy, in anno 1357."*

586. KIRKALDY, JAMES.

A fess between two estoiles in chief and a crescent in base.

" JACQUES KYRKELDE."—*Appended to "Covenants from Norman Lesly, relating to the proposed Marriage between the Queen of Scots and King Edward VI., 15th March 1546.—H.M. Record Office.*

587. KIRKLAND, DAVID.

On a bend, three cinquefoils ; in the dexter base point a leaf.

" [◁] DAVID KYRKLAND," A.D. 1547.—*Glasgow College Charters.*

588. KIRKPATRICK, JOHN.

A saltire and chief.

" S' JEHAN DE KIRKPATRIC." Detached Seal.—*H.M. Record Office.*

589. KIRKPATRICK, THOMAS.

A fleur-de-lis, not on a shield.—*Appended to " Concordia," by the Commissioners for Governing the Marches, 12th July 1429.—H.M. Record Office.*

590. KYD, ANDREW.

Probably only a merchant's mark. A cross (?) between two saltires couped ; foliage at the top of the shield.

" S' ANDREE KYDE."—*Appended to an Instrument dated 22d November 1473. David Laing, Esq.*

591. KYNMAN, NICOLAS, OF MEGGINCH.

Two escutcheons in fess : a crescent in base. A label of three points.

" S' NICOLAII KYNMAN."—*Appended to Resignation of the lands of Megginch to William Earl of Errol, 1st September 1461.—Errol Charters.*

592. LAMB, JOHN, BAILIE, BURGESS OF EDINBURGH.

The Agnus Dei between two mullets pierced in chief, and a mascle in base : foliage at the top and sides of the shield.

" S' JOHANNES LAMBE."—*Appended to the same Instrument as No. 356.*

593. LAMBERTON, ROBERT.

A stag's head cabossed, not on a shield, between the antlers a cross, and in lower part of the seal two fleurs-de-lis, a crescent, and a star.

" S' ROBERTI LAMBERTON."—*Appended to Charter of his lands of Eyton, Eymouth, Coldingham, and Flemington, to William Stute, of Berwick, A.D. 1336.—Dean and Chapter of Durham.*

594. LAMBERTON, SIMON.

Oval shape. Within a Gothic niche the Virgin and Child ; in the base a monk praying.

" S' SIMONIS D. LAMBRETOUSE CLICI." Detached Seal.—*H.M. Record Office.*

595. LANDALE, FIRRIN.

A stag's head cabossed; between the antlers a shield bearing a mullet of six points in centre of an orle.

" s' FIRRIN (?) DE LANDALES."—*Appended to the same Homage as No. 48, A.D. 1292.* —*H.M. Record Office.*

596. LANDEL, JOHN.

An eagle displayed, within an orle; tracery around the shield.

" SIGILL. JOHE DE LANDELE." Detached Seal.—*H.M. Record Office.*

597. LANDEL, THOMAS.

Shield-shaped seal. A wolf (?) rampant, not on a shield.

" s' THOME DE LANDELAS." Detached Seal.—*H.M. Record Office.*

598. LANGTON, ALAN.

An antique gem. Figure of Ceres.

" s' ALANI DE LANGTWN." Detached Seal.—*H.M. Record Office.*

599. LANGTON, WALTER.

Device, not on a shield, of an eagle preying on a hare.

" SECRETUM WALTERI DE LANGETONE." Detached Seal.—*H.M. Record Office.*

600. LARDENER, ALEXANDER.

Device of a bird on a tree, a dog or lion at the foot.

" s' ALEXSANDRI DE LARD." Detached Seal.—*H.M. Record Office.*

601. LASCEL, JOHN.

Device of a hand and falcon; a mullet in the background.

" s' JOHANNIS DE LASHCEL." Detached Seal.—*H.M. Record Office.*

602. LATWYRE (?), PETER.

An antique gem. Thespis, or an actor, and a mask.

" s' PETRI DE LATWYRE."—*Appended to the same Homage as No. 48, A.D. 1292.*— *H.M. Record Office.*

603. LAUDER, ALEXANDER, BURGESS OF EDINBURGH.

A griffin segreant; foliage at the sides and top of the shield.

"S' ALEXANDRE LAUDER."—*Appended to the same Instrument as No. 90.*

604. LAUDER, GEORGE, OF THE BASS, A.M.

A griffin segreant within the royal tressure. Rose and thistle ornament around the shield.

"S' M. GEORGII LAUDER DE BAS."—*Appended to Charter of the mill of Mersingtoun, etc., in parish of Eccles, co. Berwick, to James Maitland of Lethingtoun, 21st February 1603.—David Laing, Esq.*

605. LAUDER, SIR ROBERT, OF THE BASS, LORD-JUSTICE OF SCOTLAND, ETC. ETC.

Couché; a griffin salient, within the royal tressure. Crest, on front-faced helmet, a goat or ram's head. Supporters, two lions rampant.

"S' ROBERTI DE LAWEDRE."—*Appended to an Instrument dated 16th July 1425.— In possession of David Laing, Esq.*

606. LAUDER, ROBERT, OF THAT ILK.

A griffin segreant.

"S' ROBERTI LAUDERE."—*Appended to Assignation of Lands of Hopton, in barony of Edilston, Peeblesshire, to Martin Balcasky, burgess of Peebles, and Christian his spouse, 23d May 1504.—Philliphaugh Charters.*

607. LAUDER, WILLIAM.

A griffin segreant.

"S' GULLIM LAWEDRE." Detached Seal.—*Melrose Charters.*

LAUDERDALE, EARL OF. *Vide* MAITLAND.

608. LAWSON, JAMES.

Two crescents in chief, and a mullet in base; foliage at the top and sides of the shield.

"S' JACOBI LOUSSOUN."—This is probably the seal of his father, James Lawson, A.M., who was assistant and successor of John Knox as minister of St. Giles, Edinburgh, and died in London 1584. —*Appended to the same Instrument as No. 459.*

609. LAZOUCHE, Lady ELEANOR, second Daughter and
Co-heiress of Roger de Quinci, Earl of Winchester, and Wife of . . . . Lazouche, Baron
of Ashby.

    A fine seal. Oval shape, a full-length figure of a
lady, holding in each hand a shield, the dexter
charged with ten bezants, four, three, two, and one,
the coat of Lazouche; the sinister bears a cinquefoil,
derived from Robert de Bretuil (Fitzparnel), Earl
of Leicester.

    " SIGILL. DNE ELEINE LAZOCL."—*Appended to an Instru-
ment about A.D.* 1298.—*Communicated by W.
Fraser, Esq.*

610. LEARMONTH, ANDREW.

    On a chevron, three mascles.

    " s' ANDRIE LERMONTUT."—*Appended to a Sasine,* A.D. 1520.—*St. Leonard's College
Charters.*

611. LEARMONTH, SIR JAMES, of Balcomy.

    On a chevron, three mascles. Crest, on helmet with mantling, a hand holding a
cornucopia.

    " SIGILLUM JACOBI LEARMONCHT MILITIS." Detached Seal.—*H.M. Record Office.*

612. LEARMONTH, JAMES, of Dairsie.

    On a chevron, three lozenges.

    " s' JACOBI LARMONTH."—*Appended to an Indenture dated* 1530.—*St. Andrews
Charters.*

613. LEARMONTH, PATRICK, of Dairsie.

    On a chevron, three mascles. The sides and top of shield ornamented with
foliage.

    " s' PATRICII LERMONT."—*Appended to* "Indented Charter" *by Alexander Howbroun,
chaplain of the Altar of St. Dothac, Thomas Welwood, and Jonet Geddes his
spouse, of a tenement on the north side of the* "mercat gate," *St. Andrews,* 22d
January 1554.—*Charters of St. Mary's College, St. Andrews.*

614. LEIGHTON, JOHN, of Ulshaven. (From this
family is descended the excellent and celebrated Archbishop Leighton.)

A lion rampant. Foliage at the top and sides of
the shield.

" s' JOHANIS LICHTON DE VLLSHEL."—*Appended
to a Charter by John Leighton of Ulshaven
to David Hogue, of the halflands of Licch-
ton, in the barony of Inverdovat, 7th February 1572.—Crawford Charters.*

615. LELSOL, JOHN.

A wheel ornament.

" s' JOHIS D. LELSOL."— *Appended to the same Homage as No. 48, A.D. 1292.—H.M.
Record Office.*

616. LENNOX, MALCOLM, EARL OF.

A saltire cantoned with four roses. A label of three points. The shield in centre
of elegant tracery.

" s' MALCOMI D. LENNOX."—*Appended to a fragment of a Deed, probably a Deed of
Homage, 1292.—H.M. Record Office.*

617. LENNOX, HUGH, PERPETUAL VICAR OF MONIMAIL.

A chevron between two crescents in chief, and a cinquefoil in base.

" s' DOMINI HUGO LEVENAX."—*Appended to a Charter by Hugh Lennox to Richard
Young, Rector of Lonmynton, of annual-rents from some tenements in the Venal,
St. Andrews, A.D. 1463.—St. Leonard's College, St. Andrews.*

618. LESLIE, ANDREW, MASTER OF THE HOSPITAL OF SPEY.

On a fess, between two mullets, one of the same ; inter two buckles ; tongues
fesswise.

" s' ANDRE LECLY . . . ."—*Appended to Charter by Andrew Leslie, Master of the
Hospital of Spey, with consent of the Bishop and Chapter of Moray, to W.
Cawdor, of the Mill of Nairn, 10th June 1471.—Cawdor Charters.*

619. LESLIE, CHRISTIAN, WIFE OF ALEXANDER LESLIE.

On a bend, three buckles, surmounted with a fess charged with as many of the same.
Inscription illegible.—*Appended to the Gift of an annual-rent to the Chaplains
of the Choir of the Collegiate Church of Aberdeen, A.D. 1427.—Lord Lindsay.*

620. LESLIE, JOHN, sixth (?) Earl of Rothes.

A signet, very prettily executed. Quarterly: first and fourth, on a bend, three buckles, tongues erect, for Leslie; second and third, a lion rampant for Abernethy. Above the shield a coronet, and around it the initials $_{I \cdot R}^{E \cdot}$. (John Earl of Rothes).—*From a Letter addressed to Sir Robert Kerr of Ancrum, dated 14th April 1624.—Marquis of Lothian.*

621. LESLIE, PATRICK, LORD LINDORES, second Son of Andrew, fifth Earl Rothes.

Quarterly: first and fourth, on a bend, three buckles for Leslie; second and third, a lion rampant for Abernethy (?). An escutcheon surtout, bearing a castle, for title of Lindores. A label of two points for difference. Crest, on helmet with mantlings, an angel's head with wings expanded. Supporters, two griffins segreant. Below the shield the motto " stat promissa fides." Inscription around the seal, " s' patricii domini de lundoris."—*Appended to Precept of Clare Constat to Alexander Irvine of Drum, 8th July 1615.—Communicated by Lord Lindsay.*

622. LESLIE, . . . LORD.

Quarterly: first and fourth, on a bend, three buckles; second and third, a lion rampant. Crest on helmet, a hand holding a dagger. Supporters, two griffins. " sigillum dns lesley." Detached Seal.—*H.M. Record Office.*

623. LESTALRIG (or RESTALRIG), EDWARD.

Oval shape. A horse, or lion passant, not on a shield. The inscription is much damaged.

" sigillum eadardi de lastalr."—*Appended to a Charter by Edward " fil. Petri de Lastairie Baro. regis Scotie," confirming charter of two tofts at Eyemouth and one at Leith, which Robert, son of Matilda, of Berwick, gave to the Priory of Coldingham.—Dean and Chapter of Durham.*

624. LESTALRIG (or RESTALRIG), SIMON.

An eagle with wings expanded, not on a shield. " s' simonis de lasalric." Detached Seal.—*H.M. Record Office.*

625. LETHAM, RICHARD.

A device, not on a shield. An arm issuing from the dexter side of the seal, vested in a hawking-glove, holding a falcon by its jesses. " s' ricardi de lethamni."—*Melrose Charters.*

626. LIDDEL, SIR JAMES.

Quarterly: first and fourth, on a bend, three mullets; for Liddel, second and third, a lion rampant. Inscription not deciphered."—*Appended to Treaty of Alliance between Alexander, Duke of Albany, and King Edward IV., against the King of Scots, James III., 11th February 1482.—H.M. Record Office.*

627. LIDDEL, JANET, LADY OF HALKERSTON.

On a bend, three mullets of six points, pierced.

" S' JANETE LIDDALE."—*Appended to Charter of some lands in Selkirk to Alexander Murray, 17th August 1554.—Blackbarony Charters.*

628. LIDDEL, JOHN, OF HALKERSTON.

On a bend, between an eagle's (?) head erased in chief and an otter's (?) head erased in base, a mullet.

" S' JOHANNIS LIDDEL DE HAKER."—*Appended to Charter of lands of Elibank to Sir Gideon Murray, 12th March 1594.—Blackbarony Charters.*

629. LINDSAY, SIR WALTER, KNIGHT.

A fragment of a fine seal; an eagle standing with wings displayed, not on a shield, holding a flower in its beak. In the field are letters somewhat resembling 𝕫, or it may be a monogram.

" [S' WALTERI] LINDESEI FILII WILL."—*Appended to an Indenture between Arnold Prior of Coldingham and Sir Walter Lindsay, Knight, regarding the Chapel of Lamberton; without date, but certainly as early as the twelfth century.—Dean and Chapter of Durham.*

630. LINDSAY, SIR DAVID, KNIGHT. *Plate VI. fig. 1.*

An eagle standing, wings expanded, not on a shield.

" SIGILLUM DAVID DE LIN[DESE]YA."—*Appended to Donation by Sir David Lindsay to the Abbey of Holyrood, Edinburgh, of the lands of Slepersfield, co. Peebles; not dated, but must be early in the thirteenth century.—In the possession of Captain Kennedy, H.E.I.C.S., Bath.*

This seal, as well as the preceding one, is a most interesting and satisfactory confirmation of the statement that the eagle—the heraldic charge of the Norman Limesays—was the cognizance of the Lindsays ere heraldic science had attained a definite form in this country. Subsequently they adopted the fess chequé, which, with the Lion of Abernethy, has for centuries held its place on the family shield.

o

631. LINDSAY, DAVID, LORD OF CRAWFORD.

     Couché ; a fess chequé, within a bordure charged with eight buckles. Crest, on helmet, a key ; supporters, two lions.

     "S' DAVID LINDESAY."—" Sigillum Domini Davidis Lindesey Domini de Crawford in anno 1345."—Hutton's " Sigilla," p. 177.

632. LINDSAY, ALEXANDER, SECOND EARL OF CRAWFORD. Plate I. Frontispiece, fig. 9.

     Couché. Quarterly: first and fourth, a fess chequé for Lindsay; second and third, a lion rampant, debruised with a ribbon, for Abernethy. Crest on helmet, a swan's neck issuing from a coronet. Supporters, two lions rampant, gardant, coué. Background ornamented with foliage, and behind each supporter an ostrich feather. The inscription on this fine seal is partly lost, but has evidently been "S' ALEXANDRI LYNDESAY, COMITIS DE CRAUFURDE."—Appended to Charter by the Earl, with consent of his first-born son, David, to Sir Walter Ogilvy, Lord of Lintrathen, of the lands of Halyards, etc., in Perth, 20th June 1424.—Communicated by Lord Lindsay.

633. LINDSAY, DAVID, SON OF THE ABOVE EARL ALEXANDER, AND AFTERWARDS THIRD EARL OF CRAWFORD.

     Much damaged. Quarterly, Lindsay and Abernethy, as in the last, the Lindsay quarters differenced with a label of three points. Foliage at the top of the shield. Very inferior style of work.

     "S' DAVIDH LINDESAY."—Appended to the same Instrument as the last.

634. LINDSAY, DAVID, FIFTH EARL OF CRAWFORD, CREATED DUKE OF MONTROSE 1486. Plate VI. fig. 2.

     A rod or staff of office supporting a ducal crown. At each side are the initials D. L., united with a cord and tassels. The latter part of the inscription is lost.

     "S' DAVID COMITIS CRAUFURDE . . . . ." This has probably been the official seal of the Duke as Justiciar "benorth the Forthe," and Chamberlain of Scotland.—From the Hutton Collection.

635. LINDSAY, JOHN, SIXTH EARL OF CRAWFORD, SON OF THE ABOVE EARL DAVID.

     Quarterly : first and fourth, chequé; second and third, Abernethy, as before. There can be no doubt that the fess chequé for Lindsay is intended in the first and fourth quarter, but the engraver has erroneously made the entire chequé. Crest, on helmet with mantlings, a swan's neck issuing from a coronet.

"s' JOHANNIS LINDESAY COMES."—*Appended to Precept of Sasine of Lands in Perthshire, in favour of James Lord Ogilvy of Airlie, 15th April 1506.— Communicated by Lord Lindsay.*

636. LINDSAY, DAVID, ELEVENTH EARL OF CRAWFORD. *Plate VI. fig. 3.*

Quarterly, Lindsay and Abernethy, but, contrary to usual practice, Abernethy occupies the first and fourth quarters Crest, on helmet with mantlings and a coronet, a swan passant, with a ring in its mouth; on a ribbon "PURE FORT." Supporters, two lions rampant.

"s' DAVIDIS COMITIS CRAUFURDE DNI LYNDESAY."—*Appended to Charter of Hallyards, etc., in Perthshire, to James Lord Ogilvy of Airlie, A.D. 1605.—Communicated by Lord Lindsay.*

637. LINDSAY, SIR WILLIAM, OF THE BYRES.

A perfect and pretty seal. Couché: a fess chequé; in chief three mullets. Crest on helmet, a swan's neck issuing from a coronet. Supporters, two griffins.

"s' WILLI DE LINDESAY."—*Appended to Charter by Sir William Lindsay of the Byres, of Lands in Roxburgh to Andrew Lindsay, his son; not dated, but about A.D. 1390.—Home Charters.*

638. LINDSAY, JOHN, FIRST LORD LINDSAY OF THE BYRES.

The same design as the last, and equally well executed.

"s' JOHANIS DE LYNDESAY."—*Appended to Precept of Sasine for infefting Walter de Innes in the lands of Aberkerdour, A.D. 1456.—Communicated by Lord Lindsay.*

639. LINDSAY, JOHN, FIFTH LORD LINDSAY OF THE BYRES.

Couché; to the sinister, as before. Crest, on helmet with mantling, a griffin's head. Supporters, two griffins.

"s' JOHANNIS DOMINI LYNDESAY."—*Appended to an Agreement dated 28th May 1542. —Edinburgh Charters.*

640. LINDSAY, JOHN, Tenth Lord Lindsay and Seventeenth Earl of Crawford.

    A signet, with a shield, without any external ornament whatever, except the earl's coronet above; the shield bears a fess chequé, and in chief three mullets.—*From a Letter to the Earl of Lothian, dated 2d July* (1650).—*Marquis of Lothian.*

    There is another seal of this Earl to a letter dated 21st January 1650, but it is merely the crest, a swan passant on an earl's coronet.

641. LINDSAY, JOHN, Twentieth Earl of Crawford and fourth Earl of Lindsay (of the Byres).

    Quarterly: first and fourth, counter-quartered Lindsay and Abernethy; second and third, a fess chequé, and in chief three mullets for Lindsay of the Byres. Supporters: dexter, a lion sejant; sinister, a griffin segreant. Above the shield an earl's coronet, and below it the motto, on a ribbon, "INDURE FORT."

    This very pretty example of the style of art in the early part of last century is from the original carnelian seal in the possession of Lord Lindsay.

642. LINDSAY, SIR DAVID, of Edzell, Knight.

    Quarterly: first and fourth, a fess chequé for Lindsay; second and third, a lion rampant, debruised with a ribbon, for Abernethy. The shield, ornamented with foliage, between the initials D. L... and above it a helmet but no crest, in its place a ribbon, with the motto, "DUM SPIRO SPERO."

    "S' DAVIDIS LINDESAY DE EDZELL."—*Appended to a Charter by Sir D. Lindsay to William Lockie and Isobel Ogilvy his spouse, of an annual rent of ten bolls of victual from the sunny half of the lands of Garlabank, 30th October* 1589.—*Crawford and Balcarras Charters.*

643. LINDSAY, SIR WILLIAM, Knight, of Rossy.

    Couché; a fess chequé; in dexter chief point a mullet. Crest, a swan's neck issuing from a coronet.

    "S' VILLELMI LYNDESAY DE ROSSY."—*Appended to Charter by Sir William Lindsay, of lands in Fife to David Stuart,11th May* 1423.—*Communicated by W. Fraser, Esq.*

644. LINDSAY, JOHN, of Covington.

    A fess counter componé between a mullet in the sinister chief point, and a lozenge in base.

    "S' JOHANNIS LENDICCAY," A.D. 1519. Detached Seal.—*Communicated by W. Fraser, Esq.*

645. LIVINGSTON, ALEXANDER, of DUNIPACE.

Quarterly: first and fourth, three lozenges (billets?) for Callendar; second and third, three cinquefoils, or gillyflowers, for Livingston; foliage at the sides and top of the shield.

"S' M. ALEXANDRI LEVINGSTOUN."—*Appended to a Charter by A. Livingston of some lands in Perthshire to St. Andrews.* A.D. 1442.—*St. Andrews Charters.*

646. LIVINGSTON, ALEXANDER, of DUNIPACE.

Quarterly: first and fourth, three cinquefoils for Livingston; second and third, three billets bendways for Callendar; foliage at the top and sides of the shield.

"S' M. ALEXANDRI LEVISTOINGE."—*Appended to a Charter by A. Livingston, with consent of John, his son and heir-apparent, to J. Giveirson, Prior, of some tenements in St. Andrews,* A.D. 1552.—*St. Andrews Charters.*

647. LIVINGSTON, ALEXANDER, COMMISSIONER FOR THE BARONY OF BROUGHTON.

A fine large seal. Three cinquefoils (gillyflowers).

"S' M. ALEXANDRI LEVINGSTOUN COMIS BARONIS BARONIE DE BROCHTON."—*Appended to an Instrument dated 25th February* 1592.—*Edinburgh Charters.*

648. LIVINGSTON, JOHN.

Three cinquefoils; foliage at the top and sides of the shield.

"S' JOANNIS LEVINGSTOUN."—*Appended to Charter by Eufamie Robert to George Bog, of an acre of land lying near the standing-stone without the West Port of Linlithgow, 28th November* 1584.—*Colston Charters.*

649. LIVINGSTON, JOHN, A.M., MINISTER OF THE PARISH OF ANCRUM, ROXBURGHSHIRE.

A signet, with the shield quarterly: Livingston and Callendar; above the shield are four Hebrew characters: אבנה (Ebene[zer]?).—*From a Letter to the Earl of Lothian, dated 29th July* 1648.—*Marquis of Lothian.*

650. LIVINGSTON, WILLIAM.

Three cinquefoils (gillyflowers) within a double tressure, flowered and counterflowered.

"S' W. D. LEVINGSTOUN." Much damaged.—*Appended to Instrument relating to the Ransom of King David II., 5th October* 1357.—*H.M. Record Office.*

651. LIVINGSTON, WILLIAM, LORD.

A small signet. Quarterly: first and fourth, three cinquefoils within a double tressure, fleury, counter-fleury, for Livingston; second and third, a bend between six billets for Callendar. Crest, on a helmet with mantling, a demi-savage, holding with both hands a club sinister bendwise in front; around the shield the initials $\overset{W}{\underset{L}{\cdot}}$ (William Lord Livingston), A.D. 1590.—*Morton Charters.*

652. **LIVINGSTON, SIR JAMES,** created Earl of Callendar, a.d. 1641.

> A signet. A shield quarterly: Livingstone and Callendar, as in the last; above the shield an earl's coronet. Supporters, two lions rampant gardant.—*From a Letter to the Earl of Lothian, dated 15th July 1649.—Marquis of Lothian.*

653. **LIVINGSTON, WILLIAM,** Rector of Monyabrotht (Kilsyth).

> Quarterly: Livingston and Callendar as before, except that the tressure is here *single,* not double, as in the former; foliage at the top and sides of the shield. " s' M. GULI LIVINGSTOUN REC. DE MOND.," a.d. 1609.—*Kilsyth Charters.*

654. **LOCKHART, ANDREW.**

> A fess between three fetterlocks.
> " s' ANDRE LOCCART."—*Appended to Procuratory of Resignation of the lands of Blatonnally, in Perthshire, a.d. 1410.—The late George Smythe, Esq.*

655. **LOCKHART, SIR GEORGE,** of Carnwath, Lord Advocate, a.d. 1658; Dean of Faculty, 1672; Lord President, 1685; Assassinated, a.d. 1689.

> A pretty and well-executed signet. A heart within a fetterlock; on a chief three boars' heads couped. Crest, on a helmet with mantlings, a boar's head erased.—*From a Letter to the Earl of Lauderdale, dated 29th March (1661).—David Laing, Esq.*

656. **LOGAN, GEORGE,** of Bonnington.

> Three piles in point.
> " s' GEORGII LOGANE DE BONNYTOUN."—*Appended to a Charter by G. Logan, of the Bonnington Mills to the Burgh of Edinburgh, 2d May 1617.— Edinburgh Charters.*

657. **LOGAN, SIR JOHN,** of Restalrig, Knight.

> Three piles in point; the shield ornamented with foliage.
> " s' JOHANIS LOGAN DE LESTALRIG MILES."—*Appended to Charter by Sir John Logan of a tenement at Leith to James Mackyson, 20th February 1504.—Edinburgh Charters.*

658. **LOGAN, ROBERT,** Son and Apparent Heir of Sir Robert Logan of Restalrig.

> Quarterly: first and fourth, three piles in point for Logan; second, an eagle displayed for Ramsay; third, three popingoes for Pepdie; the shield of a fanciful shape, and ornamented with foliage.
> " s' ROBERTE LOGANE DE RESTALRIC."—*Appended to Sasine of a piece of ground at Leith, called the Links, to James Creich, 31st August 1542.—Edinburgh Charters.*

659. LOGAN, ROBERT, of RESTALRIG.

Three piles in point.

"SIGILLUM ROBERTI LOGAN DE RESTALRIK."—"*Sigillum Roberti de Logan Domini de Restalrige in Lautheana, A° 1279.*"—*Hutton's "Sigilla,"* p. 188.

660. LOGAN, WALTER.

A stag's head cabossed; between the antlers a shield, bearing three piles conjoined in point.

"SIGILLUM WALTERI LOGAN." Detached Seal.—*H.M. Record Office.*

661. LOGY, THOMAS.

Two chevronells; in base a lion's face.

"S' MAGRI TOME LOGY."—*Appended to a Charter by Thomas Logy, dated A.D. 1467.* —*St. Leonard's College Charters.*

662. LONDON, WILLIAM, SUPPOSED SON OF WILLIAM DE LUNDRES, NATURAL SON OF WILLIAM THE LION, KING OF SCOTLAND.

Device, not on a shield. An eagle seizing a bird; above the back of the eagle a branch of foliage.

"S' WILLI FILL. WILLI DE LONDON."—*Brass matrix, in possession of Richard Sainthill, Esq., Cork.*

663. LONGUS, THOR.

Oval-shaped seal. A front figure of a knight (?) sitting bare-headed, and not in armour, holding with both hands a sword obliquely across his breast.

"THOR ME MITTIT AMICO."—This extremely interesting and curious seal is appended to his charter granting the Church of Edenham (built by him in honour of St. Cuthbert) and a caracute of land to St. Cuthbert and his monks, in perpetual possession. The charter is without date, but must be as early as 1098.—*Dean and Chapter of Durham.*

664. LOVELL, JOHN, SON AND APPARENT HEIR OF HENRY LOVELL OF BALUMBY.

Three piles in point; two bars wavy. The paternal shield of Lovell bears a fess, wavy, surmounting the piles. The two bars in this, below the piles, may probably be regarded as the mark of cadency assumed during the life of the bearer's father.

"s' JOHANNES LOVEL."—*Affixed to a paper dated 6th February 1557.—Lyon Office.*

665. LOVELL, JAMES, OF BALUMBY.

Three pales, surmounted by as many barrulets.

"s' JACOBI LOVELL."—*Appended to a Charter by James Lovell, senior, Treasurer of Dundee, of a dwelling house in said burgh to Robert Lovell, senior, and Robert Lovell, junior, his lawful son by Margaret Wedderburne, 13th October 1575.— Communicated by J. C. Roger, Esq.*

666. LUGAN, ANDREW.

Oval shape. A figure of St. Andrew; above his head a star; in base a monk praying. A crescent and foliage at the sides.

"SIGILL. ANDREE DE LUGAN." Detached Seal.—*H.M. Record Office.*

667. LUMSDEN, ADAM.

On a bend, two mullets.

"s' ADE DE LUMSDEN."—*Appended to Perambulation of the lands of Adam Forman. A.D. 1430.—Dean and Chapter of Durham.*

668. LUMSDEN, ALEXANDER, PARSON OF FLISK, FIFESHIRE.

On a chevron, three mullets; in base a rose.

"[s' ALEXANDRI] LUMSDEN."—*Affixed to paper, the decision of " Mag. Alexr. Lumisden," as one of the arbiters between David Berklow, Laird of Callerny, and Henry Petcarny (Pitcairn) of that ilk, regarding the exchange of some lands, 30th April 1485.—David Laing, Esq.*

669. LUMSDEN, GILBERT.

A bend sinister, engrailed, between two mullets in chief and one in base.

"s' GILBERTI DE LUMESDEN."—*Appended to the same Instrument as No. 667.*

670. LUMSDEN, THOMAS.

On a bend, two mullets of six points.

"s' THOME DE LUMSDEN."—*Appended to Inquisition of the lands of Thomas Lumsden in Coldingham, A.D. 1431.—Dean and Chapter of Durham.*

671. LUMSDEN, THOMAS, PARSON OF KINKEL.

Much damaged. On a chevron, three mullets.

"s' M. THOME LUMSDEN."—*Appended to Tack of Kynnock to George Forbes of Kynnock, A.D. 1576.—Communicated by Lord Lindsay.*

672. LUNDIE, WILLIAM, ONE OF THE BAILIES OF ST. ANDREWS.

Three pales, surmounted by a bend, charged with as many mullets. The inscription is quite illegible.—*Appended to the same Instrument as No. 583.*

673. LYON, ARCHIBALD.

A lion rampant within a double tressure, flowered and counter-flowered. Foliage at the top and sides of the shield. Very rudely executed.

" s' AERCHBALDUS LYOUNE." A.D. 1577. Detached Seal.—*H.M. General Register House.*

674. LYON, DAVID, OF BRAKY.

A lion rampant within the royal tressure.

" s' DAVID LION."—*Appended to Reversion of an annual-cent from the lands of Kia blathmont, co. Forfar, 1st January 1506.—Crawford Charters.*

675. MACALLISTER, JOHN.

A tree growing on a mount between a sinister hand apaumé, fesswise, couped, on the dexter, and a galley on the sinister.

" s' JOANNES MVELORT."—*Appended to Reversion by John Murdolach M'Allister, Captain of Clanronald and Allan M'Ean Murdolach M'Allister, his son and apparent heir, of lands in Glenelg, to Hugh, Lord Fraser of Lovat, 8th July 1572.—Lovat Charters.*

676. MACALLISTER, ALLAN, SON OF THE ABOVE JOHN MACALLISTER.

Exactly the same charges as in the last.

" s' ALLAN MAKVICALSTIR."—*Appended to the same Instrument as the last.* Both these seals are rudely executed, but are in good condition, and extremely interesting, as being early examples of the heraldry of the Highland clans.

677. MACGILL, THOMAS, CANON OF DUNKELD AND PREBENDARY OF FOXGHORT.

Much damaged. Three birds (martlets?) passant.

"s' THOMAS MACHILL."—*Appended to a Charter by Thomas Maegill, with consent of the Bishop (Robert Crichton) and Chapter of Dunkeld, to John Douglas, natural son of Mr. W. Douglas, of the lands of Parkwood, etc., in the barony of Dunkeld and shire of Perth, 6th April 1569.—David Laing, Esq.*

678. MACGIL (or M'GIL), HELEN, WIFE OF SIMON CERRIE.

Three birds (doves?) passant; a mullet in centre, probably as a difference; foliage at the top and sides of the shield.

"s' HELENE MAGIL."—*Appended to the same Instrument as No. 266.—Edinburgh Charters.*

679. MAKDOWEL (or M'DOWAL), SIR WILLIAM, OF NEWFLANT, ENGLISH RESIDENT IN THE NETHERLANDS.

A small signet. Per pale dexter, a lion rampant contourné; in chief a rose or mullet; sinister, per fess, in the first, a stag's head cabossed; in the second a heart. Crest, on helmet with mantlings, a demi-lion rampant, crowned, issuing from a coronet, between the initials w. M.—*From a Letter to the Earl of Lothian, dated Haig (or the Hague), 9th April 1653.—Marquis of Lothian.*

680. MACKORKADIL, or M'CORQUODELL, DUNCAN.

A demi-stag erased passant.

"s' DUNCANI MAKCORKATILL." A.D. 1556.—*Argyle Charters.*

681. MACLEOD, ALEXANDER, OF DUNVEGAN.

A stag's head erased; a base chequé.

"s' ALEXANDRI . . . . ."—*Appended to Charter by Alexander M'Leod de Dunvegan of the lands of Ballalraid to John M'Ane M'Leod, natural son of John M'Leod in Megnes, 29th June 1542.—Lovat Charters.*

682. MAITLAND, SIR RICHARD, OF LETHINGTON, KNIGHT, LORD PRIVY SEAL, ETC., IN HONOUR OF WHOM THE MAITLAND CLUB WAS FOUNDED.—*Plate VII. fig. 11.*

Couché; a lion rampant. Crest, on helmet with mantlings, a lion's head erased.

"s' RICHARDUS MAITLAND."—*Appended to Charter by Sir Richard Maitland and Mr. John Spence of Condy, Commissioners of the King, specially constituted under the Great Seal for receiving resignations of lands of vacant benefices, confirming to George Home of Eddrem and Jonette Balfour, his affianced spouse, the lands of Eddrem, in the barony of Coldingham, co. Berwick, 29th July 1566,—Communicated by George Logan, Esq.*

683. MAITLAND, JOHN, second Earl of Lauderdale, created Duke of Lauderdale, 1672.

A small signet. A lion rampant within the royal tressure. Crest, on helmet with mantlings, a lion sejant affronté, holding in the dexter paw a sword, and in the sinister a fleur-de-lis. It may be mentioned that the lion on the shield of Maitland has always been carried dechaussé, but there is certainly no indication of it in either of these seals.—*From a Letter to the Earl of Lothian (signed as Sir John Maitland), dated 31st October 1644.—Marquis of Lothian.*

684. MALBRIT, KILCRIST.

An eagle, wings expanded, not on a shield.

" s' KILCRIST MALBRIT."—*Appended to the same Homage as No. 48, A.D. 1292.— H.M. Record Office.*

685 MALEWYN, ROBERT.

A lion rampant, not on a shield.

" s' ROBT . . . MALEWYN."—*Appended to the same Homage as the last.—H.M. Record Office.*

686. MALINEST, JOHN.

Device of a hand and arm holding a falcon.

" s' JOHANNIS DE MALINEIST."—*Appended to the same Homage as the last.—H.M. Record Office.*

687. MALVIN (or MELVILE), WILLIAM.

A merchant's mark on a shield.

" s' WILELMI MALVIN."—*Appended to Precept or Instrument of Infeftment, dated 28th January 1480.—In possession of David Laing, Esq.*

688. MANDERSTON, JOHN.

A chevron between three birds (?) passant.

" s' JOHANNES DE MANDERSTON."—*Appended to Inquisition on the death of G. Wardbar of the lands of West Reaton, Aymouth, and Nether Ayton, A.D. 1426.— Dean and Chapter of Durham.*

MARCH, EARL of. *Vide* DUNBAR.

689. MARR, DONALD, EARL of.

An antique gem, with Arabic or Cufic characters.

" SIGILLUM SECRETI."—*Appended to the Homage Deed of Donald Earl of Marr, A.D. 1295.—H.M. Record Office.*

690. MARR, SIR DONALD, Knight, Son of Gratney
Earl of Marr, and Christian Bruce, Sister of
King Robert Bruce.

A bend, charged on the upper part with a mullet,
between six cross-crosslets fitchée.

"s' donaldi de marre militis."—*From General Hatton's Collection.*

691. MARR, SIR DONALD, Knight.

Semé of cross-crosslets fitchée, a bend, in sinister chief point a mullet.

"s' donaldi de marre militis." Detached Seal.—*H.M. Record Office.*

692. MARR, DONALD, Earl of.

A fourth seal of the same person as the last, and a remarkably fine design, which
was subsequently adopted by his son Thomas Earl of Marr. See Nos. 565
and 566 of "*Descriptive Catalogue of Scottish Seals.*" An armed knight on
horseback, with a sword in his right hand, and on his left arm a shield charged
with a bend between six cross-crosslets fitchée; the caparisons of the horse
have also the same. On the top of the helmet is a crest resembling a pair of
wings, on which are the bend and cross-crosslets; at the dexter side of the seal
is a tree with a bird in the upper branches. The whole background is richly
ornamented with foliage.

"sigillum donaldi comitis de marre."

693. COUNTER SEAL of the preceding.

A bend between six cross-crosslets fitchée. The shield is suspended from a tree
behind, and the whole surrounded by tracery, the dexter and sinister spaces of
which are occupied by a demi-lion supporting the shield.

"otra s' (*contra sigillum*) donaldi coitis d. marre." Early part of fourteenth
century; the Earl died a.d. 1332. Detached Seal.—*H.M. Record Office.*

694. MARR, DUNCAN, Bailie of Aberdeen.

On a fess, between two boars' heads in chief and a heart in base, three buckles.
Helmet with mantling, but apparently no crest.

"s' duncane marre."a.d. 1518.—*From General Hutton's Collection.*

695. MARR, WALTER.

A chevron between two wolves'(?) heads couped in chief, and a human heart in base.

"s' dni valteri mar."—*Appended to an Indenture dated 8th January 1536.—
St. Andrews Charters.*

696. MARSHALL, JOHN.

Semé of cross-crosslets fitchée, three horse-shoes.

"s' JOHANNIS MARESCALLI."—*Appended to the same Homage as No. 48, A.D. 1292.— H.M. Record Office.*

697. MARSHALL, FERGUS.

A horse-shoe, not on a shield.

"s' FERGUSI DE MARESCALLI."—*Appended to the same Homage as the last.—H.M. Record Office.*

698. MARSHALL, JOHN.

Three horse-shoes.

"s' JOHANNIS MARESCALLI." Detached Seal.—*H.M. Record Office.*

699. MASTERTON, JOHN, PORTIONER IN BROUGHTON.

Three boars' heads erased.

"s' JOHANNIS MATHESOUN."—*Appended to a Charter of a Mill at the Water of Leith, 25th July 1563.—Edinburgh Charters.*

700. MASTERTON, ROBERT, OF BAD.

Per pale dexter, an eagle displayed, sinister, per fess, in chief a crescent, in base a chevron (or it may be blazoned, a chevron, and on a chief a crescent; some families of the name certainly carry a chief.) In place of the usual inscription are the initials s · R · M · (Sigillum Roberti Masterton).—*Appended to Resignation of the lands of Keir, in the lordship of Culross and sheriffdom of Perth, in favour of Edward Broun, 16th January 1588.—Principal Campbell, Aberdeen.*

701. MASTERTON, WILLIAM.

A lion rampant, not on a shield.

"s' WILLI DE MAISTERTUN."—*Appended to Homage Deed, 29th August 1295.—H.M. Record Office.*

702. MATHESON, MARGARET.

Three boars' heads erased (?).

"s' MARGARETE MATHESON."—*Appended to a Charter by Margaret Matheson and John Tok, of a tenement at Leith to the City of Edinburgh, 1601.—Edinburgh Charters.*

703. MAULE, THOMAS, Son and apparent Heir to Robert Maule of Panmure.
Six escallop shells, two, two, and two. Very rudely executed.
" s' thomas maul."—*Affixed to Paper dated* 1558.—*In Lyon Office.*

704. MAXWELL, DONALD.
Oval shape. A hunting-horn, in the loop of the string a squirrel, not on a
shield.
" s' dovenaldi mac[es]oul."   Detached Seal.—*H.M. Record Office.*

705. MAXWELL, EDWARD, of Lochrutton.
A saltire, in base are three points or dots; one and two, probably as a cadency.
" s' edcardi maxwell."—*Appended to Reversion of some lands in Kirkcudbright.
8th February* 1540.—*Communicated by W. Fraser, Esq.*

706. MAXWELL, EDWARD, Son of the preceding Edward Maxwell.
A saltire between a crescent in chief and a mullet in base.
" s' edvardi maxwell."—*Appended to the same Instrument as the last.*

707. MAXWELL, JOHN.
A device of an eagle preying on a bird.
" s' johan de macseywelle."—*Appended to the same Homage as
No. 48, a.d.* 1292.—*H.M. Record Office.*

708. MAXWELL, JOHN, Son of Herbert Maxwell.
Device of a rabbit feeding.
" s' johis fil. herbti d. maceswel."—*Appended to the same Homage as the last.*

709. MAXWELL, JOHN, of Calderwood.
Quarterly: first and fourth, a saltire within a bordure,
for Maxwell; second and third, a bend, for Denniston.
" johannes de maxewelle dn. de calderwode."—*Appended to Confirmation Charter of the lands of
Hylton, in Berwick, to Adam Blackader, 3d February* 1470.—*Communicated by W. Fraser, Esq.*

710. MAXWELL, PATRICK, of Newark.

A saltire; in chief a gemmed ring, and in base a duck (?); foliage at top and sides of the shield.

"s' patrich maxwel de newerk."—*Appended to Reversion of the lands of Fasline and Bardisland, in earldom of Lennox and sheriffdom of Dumbarton, to Matthew Earl Lennox, 31st July 1532.—Communicated by W. Fraser, Esq.*

711. MAXWELL, PATRICK, of Dargavel.

A saltire; in base canton a stag's head.

"s' patrici maxvel."—*Appended to Resignation in favour of himself, 25th November 1605.—J. H. Hall Maxwell of Dargavel, Esq.*

712. MAXWELL, ROBERT.

A saltire.

"s' robarti maxwel." Detached Seal.—*H.M. Record Office.*

713. MAXWELL, ROBERT, of Calderwood.

Quarterly: first and fourth, a saltire within a bordure counter componé; second and third, a bend sinister, for Denniston. The bend being here placed sinisterwise must be ascribed to the carelessness of the engraver; the bend dexter is the proper charge.

"s' roberti maxwell."—*Appended to a Charter by Robert Maxwell of Calderwood, lord of the barony of Maldisler, to John Weir of Weightshaw, of the lands of Weightshaw, Lanarkshire, A.D. 1508.—In possession of David Laing, Esq.*

714. MAXWELL, ROBERT, Lond. *Plate VIII. fig. 9.*

A saltire; foliage at the top and sides of the shield.

"s' robarti domini maxwel."—*Appended to Obligation to the Friars Minors of Dumfries, 27th June 1551.—Communicated by W. Fraser, Esq.*

715. MELDRUM, ALEXANDER, of Segie.

A fox (lion?) rampant, no doubt intended for an otter; in dexter chief an ermine spot, and in sinister chief a mullet of six points.

"s' alexandri meldrum."—*Appended to an Instrument of Sasine, A.D. 1449.—St. Leonard's Charters.*

716. MELDRUM, DAVID, Canon of Dunkeld, and Official of St. Andrews.
    An otter rampant, crowned, contourné; in the dexter fess a mullet of six points.
    "s' DAVID MELDRUM."—*Appended to Charter by D. Meldrum, of some property in
    St. Andrews to God, and Altar of the Blessed Virgin at St. Andrews, 29th
    January 1495.—St. Andrews Charters.*

717. MELDRUM, WILLIAM, of Fyvie.
    Per pale, dexter per fess, in chief three pallets, the ancient
    coat of Meldrum, at least the same charge is on the
    seal of William Meldrum, appended to his homage,
    A.D. 1292; second, three unicorns' heads couped, for
    Preston; sinister, an otter rampant for Meldrum.
    "s' VILHELMI DE MELDRU."—*Appended to Precept of
    Sasine in favour of James Innes of "yat ilk," A.D.
    1468.—Communicated by Lord Lindsay.*

718. MELVILLE, GEORGE.
    On a fess, between three cushions, two mullets.
    "s' GEORGII MALVIN."—*Appended to an Instrument dated 22d November 1473.—
    David Laing, Esq.*

719. MELVILLE, JAMES.
    Rather damaged, but apparently a garb surmounted by a fess.
    "s' JACOBI MALEVIL." (?)   Detached Seal.—*H.M. Record Office.*

720. MELVILLE, NICOLAS.
    A fess between three crescents.
    "s' NICOLAI MALEVIL."   Detached Seal.—*Communicated by Mr. Thomas, Perth.*

721. MELVILLE, ROBERT.
    Three cushions.
    "s' ROBERTI MALVELL."—*Appended to the same Instrument as No. 715.*

722. MENTEITH, JOHN.
    A fess chequé, surmounted by a bend.
    " SIGILL. JOHANNIS MENETET."   Detached Seal.—*H.M. Record Office.*

723. MENTEITH, PATRICK, of Rotho.

A bend chequé; in sinister chief point a crescent.

" s' PATRICI MENTEITH DE RA."—*Appended to Precept of Sasine of lands in " rille" of Rotho, in Renfrew, in favour of Robert Sempill, grandson of William Sempill, 17th December 1535.—Baberton Charters.*

724. MENTEITH, WILLIAM.

Quarterly: first and fourth, a bend chequé, for Menteith; second and third, on a bend, three buckles, for Stirling of Cadder. Crest, on helmet, a swan's neck.

" s' WILLMI DE METHT "(?), A.D. 1496.—*Argyle Charters.*

725. MENTEITH, WILLIAM, of West Cars.

Couché. Quarterly: first and fourth, a bend chequé, for Menteith; second and third, on a bend engrailed, three buckles, for Stirling of Cadder. Crest, on helmet, a swan's neck issuing from a coronet.

" s' WILL. DE METEITH."—*Appended to Precept of Clare Constat to Andrew Stirling of Cadder as heir to his father, William Stirling, 25th April 1517.—Keir Charters.*

726. MENZIES, THOMAS.

Ermine; a chief.

" s' THOME MENZES."—*Appended to Precept of Clare Constat in favour of John Cawdor (Campbell) as heir to his father, Archibald Cawdor, in the lands of Geddes and Rait, 21st February 1555.—Cawdor Charters.*

727. MENZIES, THOMAS, of Pitfoddels, Provost of Aberdeen.

Ermine; a chief.

" s' THOME MENGZIES D. PITFODDLIS."—*Appended to Charter by Thomas Menzies of Pitfoddels, and a salmon-fishing on the Dee, to his son, Gilbert Menzies of Cowlie, 14th November 1573.—Communicated by Lord Lindsay.*

728. MERCER, ANDREW, of Inchbreckie.

On a fess, between two crosses patée fitchée in chief, and a mullet in base, three bezants (?).

" s' ANDREE MERCER," A.D. 1455.—*Athol Charters.*

Q

729. MERCER, ROBERT, of Innerpeffrey.

A chevron between three mullets; in chief as many ermine
spots.

"s' robarti mercer."—*Appended to Procuratory for
Resignation of the lands of Easter Dowlany and
Wester Dowlany, in the earldom of Strathern and
sheriffdom of Perth, 4th July 1465.—Græme R.
Mercer, Esq. of Gorthy.*

730. METHVEN (or MEFFEN), HENRY, Bailie of St. Andrews.

On a chevron, ensigned with a cross between the initials H. M. in chief, and a human
heart in base, a crescent.

"s' m. henrici meffen."—*Appended to the same Instrument as No. 612.*

METHVEN, STUART, LORD.   *Vide* Stuart.

731. METHVYN, JOHN.

Three eagle's heads (?).   The shield in front of an eagle.

"sigillum johannis de methvyn."—*Appended to Indenture " Trengarum Seseie,"
14th August 1451.—H.M. Record Office.*

732. MEURIST, ROWISON.

A fess, between two mullets in chief and a crescent in base.

"s' m. rowison de meurist."—*Appended to a Charter dated 17th April 1628.—
Musselburgh Charters.*

733. MICHALE (or MITCHELL ?), THOMAS, one of the Bailies of St. Andrews.

An eagle displayed.

"s' thome michale."—*Appended to an Instrument of Sasine dated A.D. 1455.—
University of St. Andrews Charters.*

734. MIDDLETON, JOHN, created Earl of Middleton 1660.   *Plate I., Frontispiece,
fig. 4.*

Per fess.   A lion rampant within a double tressure, flowered and counter-flowered
with fleurs-de-lis.   Crest, on full-faced helmet with mantlings, above an earl's
coronet, a lion rampant on the top of a tower embattled.   Supporters, two
eagles.   Motto, on ribbon below the shield, " fortis in arduis."—*From the
original silver matrix in possession of the Earl of Strathmore.*

735. MONTFORD, JOHN DE.
A device of an eagle preying on a bird. Inscription not legible.—*Appended to the Homage Deed of John de Montford, 20th July 1295.—H.M. Record Office.*

736. MONTGOMERY, JOHN, LORD OF EAGLESHAME. *Plate I., Frontispiece, fig. 8.*
Conché; an amulet gemmed between three fleurs-de-lis. Crest, issuing from a coronet above a helmet, a savage's head (or an aged head) affronté. Background foliated.
"s' JOHANNIS DE MONGUMRIE."—*Appended to Charter to William Blakeforde, 8th October 1392.—Eglinton Charters.*

737. MONTGOMERY, HUGH, LORD, CREATED EARL OF EGLINTON, A.D. 1503.
Much damaged. Quarterly: first and fourth, three fleurs-de-lis, for Montgomery; second and third, three amulets gemmed, for Eglinton. Foliage at the top and sides of the shield.
"s' HUGO COMIT. DE EGLINTON."—*Appended to Precept of Sasine, 19th April 1529. Eglinton Charters.*

MONTROSE, EARL OF. *Vide* GRAHAM.

738. MONYPENY, DAVID, OF PITMILLY.
Three cross-crosslets, issuing from as many crescents.
"s' DAVID MONIPENY OF PETMYLL."—*Appended to Covenants from Norman Leslie, relating to the proposed marriage between the Queen of Scots and King Edward VI., 15th March 1546.—H.M. Record Office.*

739. MONYPENY, JOHN, OF PITMILLY. *Plate VII. fig. 3.*
Three cross-crosslets, fitchée, issuing from as many crescents. The shield within a triangle surmounting a trefoil.
"s' JOHANIS MONIPENI."—*Appended to the same Instrument as No. 490.*

740. MONYPENY, JOHN, ONE OF THE BAILIES OF ST. ANDREWS.
A cross-crosslet issuing from a crescent.
"s' JOHANIS MONEPENYI."—*Appended to an Instrument of Sasine dated 1496.—St. Leonard's College Charters.*

741. MONYPENY, MUNGO (or KENTIGERN),
       DEAN OF ROSS.

Two cross-crosslets, fitchée, issuing from as many
crescents; in base a rose.

"S' KINTIGERNI MONIPENI DECANI ROSSE."—*Ap-
peneded to Letter of Obligation by Mungo
Monypeny, Dean of Ross, with consent of
the Chapter and Convent of the Cathedral of
Ross, to his cousin, John Monypeny, granting
him twenty pounds Scots from the church and
parish of Arthuirscheri (Arderseir), in sheriffdom of Inverness, 30th May
1573.—Cawdor Charters.*

742. MONYPENY, PETER.

Three cross-crosslets issuing from as many crescents; in chief a mullet as a
difference.

"S' PETRI MONEPENE."—*Appended to a Charter by Peter Monypeny to Richard
Young, Rector of Lamynton, of an annual rent of tenements in St. Andrews, 4th
July 1483.—St. Leonard's College Charters.*

743. MONYPENY, THOMAS, of KYNKEL.

Couché; a chevron, between three cross-crosslets issuing from
as many crescents. Crest, on helmet, an eagle (?). A
very pretty design.

"S' TOME MONEPANI D' KING."—*Appended to Charter by
James, Prior of St. Andrews, with consent of the
Chapter, to Nicolas Sytchel, of a tenement in the South
Street, St. Andrews, 17th January 1415.—St. Leonard's
College Charters.*

744. MONYPENY, THOMAS.

Couché; three cross-crosslets issuing from as many crescents. Crest, on helmet, a
stag's head. At each side of the shield is a tree growing from a mount.

"SIGILLUM THOME MONIPENE"—*Appended to the same Instrument as No. 715.*

745. MONYPENNY, SIR WILLIAM, KNIGHT. *Plate VII. fig. 6.*

The same charges as in the last. The inscription is rather broken, but appears to
be "S' WILLMI MONIPENI," A.D. 1421.—*Kilsyth Charters.*

MORAY, EARL of. *Vide* DUNBAR.

746. MOUBRAY, NIGEL (DE ALBINI), ANCESTOR OF THE MOWBRAYS OF BARNBOUGLE.

An interesting and good example of the period. The design is equestrian, a knight riding to the sinister, armed in chain or ring mail, conical helmet with nasal (?), sword in right hand, the interior side of the shield on left arm is seen.

" SIGILLUM NIGELLI DE ALBINNEIO."—*Appended to a Charter constituting his brother William his heir, in order to make over divers lands to several religious bodies, and amongst them Durham,* A.D. 1130.—*Dean and Chapter of Durham.*

747. MOUBRAY, ROGER, SON OF THE ABOVE NIGEL.

Same design as that of Nigel, but in this example the exterior of the shield is seen, the umbo bold and ornamented, and rudely formed cross-crosslets cover the remaining part, and also the mail of the knight; from the wrist of the sword-arm some drapery is flowing, perhaps a part of the skirt of the tunic.

" SIGILLUM ROGERI DE MOUBRAIE."—*Circa* 1150.—*Dean and Chapter of Durham.*

748. MOUBRAY, ROGER. THE SAME PERSON AS THE LAST.

Similar to the last. The mail, however, is decidedly chain or ring mail, with coif instead of helmet, and the shield has the umbo only. Same inscription.—*Dean and Chapter of Durham.*

749. MOUBRAY, ROGER. SAME PERSON.

Nearly the same as the last. The umbo on shield is not so prominent, and there are apparently upright and horizontal bars, as well as a border on the shield. Same inscription.—*Circa* 1180.—*Dean and Chapter of Durham.*

750. MOUBRAY, NIGEL. SON OF THE ABOVE ROGER MOUBRAY.

Similar design. The horizontal and perpendicular bars of the former shields here assume most distinctly the form of that doubtful and much-disputed charge, the carbuncle or escarbuncle, bearing a close resemblance to the fine and well-known example on the shield of Geoffrey de Mandeville, in the Temple Church, the date of which probably coincides with this.

" SIGILLUM NIGELLI DE MOUBRAIE."—*Detached Seal.—In the Bodleian Library, Oxon.*

751. MOUBRAY, JOHN, OF BARNBOUGLE.

A lion rampant crowned.

" S' JOANNES DE BARNB.," A.D. 1596.—*Cockairnie Charters.*

752. MOUSEE, GILBERT.

    A wheel ornament, or star of eight points.

    " s' GILBTI MOUSEE."—*Appended to the same Homage as No. 48*, A.D. 1296.—*H.M. Record Office.*

753. MOWAT, JAMES, CHAPLAIN OF THE HOLY CROSS OF ST. GILES, EDINBURGH.

    A small signet. A lion rampant; above the shield the initials I. M., and a ribbon, on which is a motto not legible.—*Appended to the same Instrument as No. 236.*

754. MOWAT, JAMES, THE SAME PERSON AS THE LAST.

    Two mullets, and a chief, charged with as many pallets, and a lion rampant, the latter on the sinister.

    " s' JACOBI MOWAT."—*Appended to Charter by James Mowat to the Burgh of Edinburgh, of the piece of waste ground whereon the Chapel of the Holy Cross was built, 29th March* 1619.—*Edinburgh Charters.*

755. MUNRO, ROBERT, VICAR OF URQUHART, ROSS.

    An eagle's head erased.

    " s' ROBARTI MUNRO."—*Appended to a Tack by Robert Munro, Vicar of Urquhart, in Ross, with consent of the Dean and Canons of the Cathedral Kirk of Ross, as representing the Chapter thereof, to John Campbell of Cawdor, the* " *teind salmon-fishing of the year, of Alcaig, in the barony of Ferrintosh, for payment of* 40 *shillings yearly.*" 20th *November* 1579.—*Cawdor Charters.*

756. MURE, RICHARD.

    An antique gem. A faun with a bunch of grapes and a club; at his feet a vase.

    " s' RECARDI MURE DE ROREBURA."—*Appended to the same Homage as No. 48*, A.D. 1292.—*H.M. Record Office.*

757. MURE, WILLIAM.

    A fleur-de-lis, not on a shield.

    " s' WILL. DE MURE."—*Appended to the same Homage as the last.*

758. MURE, WILLIAM, OF ABERCORN.

    Much injured. On a fess, three mullets. Crest on helmet . . Supporters, two savages.

    " s' WILL. MORE DNI DE ABYRCORNE."—*Appended to appointment of Plenipotentiaries, 26th September* 1357.—*H.M. Record Office.*

759. MURE, WILLIAM, of Abercorn.
> On a fess, three mullets. The shield in centre of elegant pointed tracery.
>> " s' willelmi more."—*Appended to Resignation of the lands of Corstorphin and Drylaw, 20th November 1397.— Sir W. H. Dick-Cunyngham of Prestonfield, Bart.*

760. MURRAY, SIR ANDREW, Knight.
> Three mullets within a bordure charged with bezants.
>> " s' andree de moravia militis."—*Appended to an Instrument relating to the Ransom of King David II., 5th October 1357.—H.M. Record Office.*

761. MURRAY, ARCHIBALD, of Darnhall (or Blackbarony).
> A fetterlock, so rudely executed as to have the appearance of a barrel, and on a chief three mullets. Rose and thistle ornament at the top and sides of the shield.
>> " s' d. archi murray de darnehalm."—*Appended to Resignation of Powerhow, etc., in Haddington, to Sir Gideon Murray of Elibank, 15th July 1617.—Balbirnie of Charters.*

762. MURRAY, EGIDIA, Lady of Colbane.
> Three mullets.
>> " s' egidie . . . moravia."—*Appended to a Discharge by Egidia Murray to Gilbert Menzies, a.d. 1438.—Communicated by Lord Lindsay.*

763. MURRAY, HUGO DE.
> A device of a hound seizing a stag. A star, and tree in the field.
>> " s' hugonis de moreve."—*Appended to Homage Deed, 1295.— H.M. Record Office.*

764. MURRAY, HUGH, of Wester Franishill.
> A saltire couped at the upper ends; in chief three mullets.
>> " s' hugonis murray."—*Appended to Reversion of Wester Franishill, 4th May 1530. —Blackbarony Charters.*

765. MURRAY, SIR MALCOLM, Knight.
> Three mullets, and a label of three points.
>> " s' malcolmi de moravia."—" *Sigillum domini Malcolmi de Moravia militis in ano. 1334."—Hutton's " Sigilla," p. 165.*

766. MURRAY, PATRICK, of FALLAHILL.

A hunting-horn stringed. On a chief three mullets. Crest, on full-faced helmet, a dexter hand vested, holding a scroll fesswise. On a ribbon below the crest, " REMEMBER."

" S' PATRICII MORRAY DE FAULAHIL."—*Appended to Reversion by Patrick Murray of the lands of Courhoip, in the barony of Romanis, in co. Peebles, 20th May 1577.*—*Blackbarony Charters.*

767. MURRAY, ROBERT, of BURNFOOT.

A cross engrailed or envecked, a pellet in upper canton, and on a chief three mullets.

" S' ROBERTI MURRAY."—*Appended to Charter of the lands of Melcamston to John Murray of Blackbarony, 5th July 1588.*—*Blackbarony Charters.*

768. MURRAY, SIR JOHN, of BLACKBARONY, KNIGHT.

A saltire engrailed, and on a chief three mullets; in base an annulet.

" SIGIL. JOHANES."—*Appended to Charter by Sir John Murray of Blackbarony to Sir Thomas Borthwick of Coplaw, and Elene Rutherford his spouse, of the lands of Culrope, in sheriffdom of Peebles, 7th June 1501.*—*Blackbarony Charters.*

769. MURRAY, SIR JOHN, of EDDLESTON, KNIGHT.

A fetterlock, and on a chief three mullets.

" S' JOANNIS MURRAY DE EDDILSTOUN MILS."—*Appended to Resignation, 21st July 1595.*—*Blackbarony Charters.*

770. MURRAY, SIR WILLIAM.

Three mullets. A rose at each side of the shield.

" S' WILLI DE MORAVIE MILIT." Detached Seal.—*H.M. Record Office.*

771. MURRAY, SIR WILLIAM, of TULLIBARDINE, KNIGHT.

Couché; three mullets within a double tressure, flowered and counter-flowered. Crest, on helmet, a peacock's head between or supported by two arms vested, couped at the elbows. Supporters, two lions rampant.

" S' WILLI MRAY TULLIBARDY."—*Appended to a Discharge of Reversion for 450 merks to Sir William Stirling of Keir, A.D. 1501.*—*Keir Charters.*

772. MURRAY, WILLIAM, of MELCAMSTON.

A fess between three mullets in chief, and a mascle in base.

" S' WILLELMI MURRAY."—*Appended to a Charter by William Murray to Kathrine Davidson and her heirs by Ninian Murray her spouse, son and heir-apparent of William Murray, of the third part of the lands of Melcamston, 26th February 1474.*—*Blackbarony Charters.*

773. MURRAY, WILLIAM.

A saltire engrailed, and on a chief three mullets.

" s' WILELMI MURRAY."—*Appended to Lease of Culrope for twenty years to Isabell Hoppar, 14th July 1551.—Blackbarony Charters.*

774. MURRAY, WILLIAM, OF ROMANOES.

A saltire engrailed, and on a chief three mullets; in base a hunting-horn stringed.

" s' WILMI MURRAY."—*Appended to Resignation of the lands of Duddingston, in Peebles, 26th November 1562.— Blackbarony Charters.*

775. MURRAY, WILLIAM, OF ROMANOES.

A saltire engrailed between three mullets, two and one in chief, and a dog's (?) head couped in base.

" s' VILIAME MURRAY." Very rude work.--*Appended to Charter of the quarter-lands of Culrope to John Murray of Blackbarony, 20th May 1577.—Blackbarony Charters.*

776. MUSCHAMP, ROBERT.

A flower or wheel ornament, not on a shield.

" s' ROBTI D. MOSCOCAM." Detached Seal.—*H.M. Record Office.*

777. MUSCHAMP, THOMAS.

Seven bees volant in saltire, not on a shield.

" SIGILLVM TOME DE MUSCHANS."—*Appended to Charter by Thomas Muschamp of a caruente of land at Bolledon to St. Cuthbert's of Durham.—Dean and Chapter of Durham.*

778. NAIRN, LAURENCE.

A chaplet.

" s' LAURENCE DE NAIRN."—*Appended to a Charter by Laurence Nairn, dated 1416—St. Leonard's College Charters.*

779. NAIRN, LAURENCE, ONE OF THE BAILIES OF ST. ANDREWS.

Three cinquefoils.

" s' LAURENCE NARN."—*Appended to a Charter by John Michael of an annual-rent of a tenement in St. Andrews to John Carmichael, 11th July 1417.—University of St. Andrews Charters.*

R

780. NARPIN, WALTER.

A wheel ornament or star of eight points, not on a shield.

" s' walteri narpin."—*Appended to the same Homage as No. 48, A.D. 1296.—H.M. Record Office.*

781. NAUGHTON, JOHN.

A device of a crescent and a star.

" s' johis de naugton."—*Appended to the same Instrument as the last.*

782. NEBRITI, GILBERT.

A cross, not on a shield.

" s' gilbert nebriti."—*Appended to the same Instrument as the last.*

783. NICOLSON, WILLIAM, of Park.

A saltire couped between the letters s. v.

" s' villelmi nicolsoun."—*Appended to the same Instrument as No. 277.*

784. NISBET, HENRY, Provost of Edinburgh.

On a chevron, between three boars' heads erased, a cinquefoil. Foliage at the top and sides of the shield.

" s' henrici neisbit."—*Appended to Charter or Grant by Henry Nisbet and Janet Brunston his spouse, of an annual-rent from a mill to the Burgh of Edinburgh, 8th June 1598.—Edinburgh Charters.*

785. NORAIS, MALCOLM.

Three bars wavy.

" s' malcolmi norais."—*Appended to same Homage as No. 48, A D. 1296.—H.M. Record Office.*

786. NORIE, JAMES, of Hildey.

A chevron between three birds, close.

" s' jacobi nore."—*Appended to Notarial Instrument, A.D. 1470.—Glammis Charters.*

787. OGIL (or OGLE), PATRICK, of Hawtreewood.

A bittern passant.

" s' patrici ogil."—*Appended to Retour of George Broun of Colston, as heir to his father, George Broun, of the lands and barony of Colston, 20th January 1538.—Colston Charters.*

788. OGILVY, ALEXANDER, Sheriff of Angus.

Couché; a lion rampant, crowned. Crest, on helmet, a lady's head veiled (?).

"s' ALEXANDRI OGYLLVYLE."—*Appended to Transumpt of "Charter by Robert Duke of Albany to Sir John Ramsay, of lands in Fife,"* A.D. 1467.—*Communicated by W. Fraser, Esq.*

789. OGILVY, ALEXANDER, "OF THAT ILK."

Couché. Quarterly: first, a lion rampant, for Inchmartin; second and third, a cross engrailed for Sinclair; fourth, a lion statant, gardant, crowned, for Ogilvy. Crest, on helmet with mantlings, a lady's head attired.

"s' ALEXANDRI OGILVY DE EODEM."—*Appended to Discharge for a sum of money to George Gordon, Constable of Badenoch, about* A.D. 1505.—*Communicated by Lord Lindsay.*

790. OGILVY, GEORGE, OF DUNLUGUS.

Quarterly: first and fourth, a lion statant, gardant, crowned, for Ogilvy; second and third, three papingoes, for House of Fast Castle.

"s' GEORGII OGIL[VI DE D]UNLUGUS."—*Appended to Charter by G. Ogilvy,* A.D. 1576. *H.M. General Register House.*

791. OGILVY, JAMES, OF CARDELL.

Quarterly: first and fourth, a lion passant; second and third, a saltire, or two clubs in saltire.

"s' JACOBUS OGYLVY."—*Appended to the same Instrument as No.* 430.

792. OGILVY, JOHN.

Couché. A lion passant; in base a human heart. Crest, on helmet with mantling, a lady's head.

"s' JOHANNIS OGILVY." Detached Seal.—*Lovat Charters.*

793. OGILVY, PATRICK, OF KILLERBREAUCH.

A lion statant; foliage at the top and sides of the shield.

"s' PATRICII OGILVI."—*Appended to Charter of part of the lands of Ard to John Ogilvy, 30th April* 1501.—*Lovat Charters.*

794. OGILVY, SIR WILLIAM, OF STRATHERNE, KNIGHT.

A lion passant, gardant, crowned.

"s' WILHELMI OGILVIE."—*Appended to Charter by Sir William Ogilvy to John Cawdor, Chanter of Ross, of the lands of Fethirty, in the Earldom of Ross and Sheriffdom of Inverness, 28th October* 1512.—*Cawdor Charters.*

795. OGILVY, GEORGE, Son of Sir Walter Ogilvy of Boyn.

Quarterly: first and fourth, a lion passant, gardant, crowned, for Ogilvy; second and third, three crescents, for Edmonston.

"s' GEORGIUS OGHILVY DE BOYNE."—*Appended to Charter by G. Ogilvy of Geddes, son and apparent heir of Sir Walter Ogilvy of Boyn, in favour of his brother-german, William Ogilvy, of the lands of Geddes and Rait, in Nairnshire, 15th October 1503.—Cawdor Charters.*

796. OGILVY, JOHN, Sheriff-Depute of Inverness.

A bend sinister; in base a lion passant, crowned.

"s' JOHANNIS OGILVI'."—*Appended to Sasine in favour of John Earl of Crawford, son of the Duke of Montrose, of the lands and barony of Strathnarne, and patronage of the church of Lundichty, or Dunlichty, co. Inverness, 22d May 1499.—Cawdor Charters.*

797. OGILVY, PATRICK.

A device of a lion passing in front of a tree, with a cinquefoil, pierced, at each side.

"s' PATRICII DE OGIVLLE."   Detached Seal.—*H.M. Record Office.*

798. OGILVY, WALTER.

A lion passant, gardant, crowned.

"s' WALTERI DE OGILVY."—*Appended to Truce for five years between King Henry VI. and King James I., 15th December 1430.—H.M. Record Office.*

799. OLIPHANT, JOHN, LORD.

Three crescents.   Crest, on helmet, an elephant's head.

"s' JOHANNIS DNI OLIPHANT."   Detached Seal.—*H.M. Record Office.*

800. OLIPHANT, WILLIAM, of Kelly.

Three crescents within a bordure engrailed or indented.

"s' WILMI OLIFANT."—*Appended to the same Instrument as No. 715.*

801. OTTERBURN, ADAM, Provost of Edinburgh, knighted A.D. 1534.

A chevron between three otters' heads erased.

"s' MAGISTRI ADE OTTIRBURN."—*Appended to Truce between Ambassadors of King James V. and King Henry VIII., 4th September 1524.—H.M. Record Office.*

802. OTTERBURN, SIR ADAM, Knight, the same as the preceding.

> A dragon or wyvern, not on a shield. No legend, but it is the seal appended by
> " Adam Otterburne" to a Convention between the Ambassadors of England
> and Scotland, relating to the Castle of Edryngtone, etc., 12th May 1534.—
> *H.M. Record Office.*

803. OTTERBURN, JOHN.

> Three otters' heads couped
> " s' MAGISTER JOHANI OTTERBURN."   Detached Seal.—*H.M. General Register House.*

804. PARK, ALEXANDER, Burgess and Dean of Guild of Edinburgh.

> A fess chequé, between three stags' heads cabossed.
> " s' ALEXANDRI PARK."—*Appended to a Reversion by Alexander Park, of the lands
> and buildings called the Salthouse, at Newhavea, to the Burgh of Edinburgh
> 18th February 1567.—Edinburgh Charters.*

805. PARKLE (or PARCLE), JAMES, Commissary of Lin-
       lithgow, and Notary-Public.

> Couché ; a chevron between three cross-crosslets fitchée.
> Crest, on helmet with mantlings, a plume of feathers.
> " JACOBI DE PARTELE."—*Appended to the same Instrument
> as No. 11.*

806. PARYS, JOHN of.

> A stag's attires ; in honour point a mullet.
> " s' JHONE OF PA[RYS]."—*Appended to Precept by James, Abbot of Dunfermline, in
> favour of Thomas Dingwall, A.D. 1505.—Seaforth Charters.*

807. PATE, ALEXANDER, Canon of the Metropolitan Church of St. Andrews.

> A carbuncle, or escarbuncle.
> " s' ALEXANDRI PATE."—*Appended to a Charter by Alexander Young, Canon of
> the Metropolitan Church of St. Andrews and Master of the College of St.
> Leonard's, with consent of the Chapter of the same, to Alexander Pate and
> Janet Nicolson his spouse, A.D. 1528.—University of St. Andrews Charters.*

808. PATRICK, Son of Edgar.

> A dragon passant to the sinister, not on a shield.
> " SIGILL. [PATRICII FILII ED]GARI."—*Appended to Charter of Edulingham Church
> to the Monks of St. Cuthbert's at Durham.—Dean and Chapter of Durham.*

809. PAXTON, DAVID.

Oval-shaped seal. An eagle displayed, not on a shield.

"SIGILLUM DAVID DE PAXTON."—*Appended to Charter of half a caracute of land in Yarford, one acre in Pollox, etc., to Coldingham.—Dean and Chapter of Durham.*

810. PAXTON, JOHN.

A cross between four ermine spots (?).

"S' JOANNE DE PAXTON."—*Appended to Perambulation of the lands of Adam Forman, A.D. 1430.—Dean and Chapter of Durham.*

811. PAXTON, JOHN.

A lion rampant; foliage at the top and sides of the shield.

"S' JOANNES [DE PAXTON]."—*Appended to Perambulation of Brockholes, A.D. 1431. —Dean and Chapter of Durham.*

812. PAXTON, WILLIAM.

Oval shape. A pelican feeding her young in a nest, not on a shield.

"S' WILLELMI DE PAXTON."—*Appended to a Document relating to fishings on the Tweed, A.D. 1250.—Dean and Chapter of Durham.*

813. PEPIDE, NICOLAS.

Oval. A bird and lion, not on a shield.

"SIGILLUM NICOLAS PAPEDI."—*Appended to Charter by Lady Matilda de Leya, of her tithes of the mill of Ancroft to the Church of Norhamshire and Islandshire. —Dean and Chapter of Durham.*

814. PINCHEST, MATTHEW, BURGESS OF ABERDEEN, AND BROTHER OF THE CARMELITES.

A crescent, and a chief, bearing on the dexter a mullet of six points, and on the sinister a cross pattée.

"S' MATTHEUS PINCHES."—*Appended to Charter of some lands in Aberdeen to the Prior and Convent of the Carmelites, 8th May 1405.—Marischal College, Aberdeen.*

815. PITCAIRN, ROBERT, ARCH-DEAN OF ST. ANDREWS AND PRINCIPAL COMMENDATOR OF DUNFERMLINE, ETC. ETC. DIED 1584. *Plate* VII. *fig.* 9.

Quarterly: first and fourth, a mascle, for Pitcairn; second and third, an eagle displayed, for Ramsay. Foliage at the top and sides of the shield.

"S' ROBERTI PITCARN, COMEDATARII DE DUNFERMLING."—*Appended to "Letter of reposition and ratification of the lands of Lymekillhill alias Langbank, and of Terrace medow, granted by the Commendator of Dunfirmling to James Murray of Perdeuis," 22d November 1578.—Elgin Charters.*

816. PLEMING, JOHN.

A bend; in sinister chief a mullet of six points. At the sides of the shield two lizards, and at the top foliage.

"s' JOHANNIS PLEMING." Detached Seal.—*Melrose Charters.*

817. POLLOCK, JOHN.

A saltire, with a lion (?) dormant in chief, and three hunting-horns, stringed, in the remaining cantons.

" SIGILLUM JOHANNIS POLLOK."—*Appended to Charter by John Louch, or Pollock, 21st April 1455.—St. Salvator's Charters.*

818. PRATT, JOHN.

A fess between three birds contourné.

" s' DNI JOHANNIS PRAT."—*Appended to Charter by Sir John Pratt, Chaplain of the Altar of Sancte Crucis, etc., in the Church of St. Nicolas, Aberdeen, A.D. 1505. —From General Hutton's Collection.*

819. PRENDERGAST, ADAM.

A lion passant, sinister, not on a shield.

" s' ADE DE PRENDERGEST."—*Appended to Resignation by Sybilla, daughter of Maurice de Aiton, of the superiority of her lands at Ayton, to Coldingham, A.D. 1246.—Dean and Chapter of Durham.*

820. PRENDERGAST, ADAM.

A lion (dog?) passant, not on a shield.

" SIGIL. ADE FIL. HENRICI DNI D. PR. DEG."—*Appended to a Quitclaim to Alexander, " Elemosinair de Coldingham."—Dean and Chapter of Durham.*

821. PRENDERGAST, HENRY.

A fleur-de-lis, not on a shield.

" SIGILLUM HENRICI DE PRENDERGUST."—*Appended to Charter of Vendition from two Monks of some lands to the Convent of Coldingham.—Dean and Chapter of Durham.*

822. PRENDERGAST, HENRY.

Ermine; three bars and a canton bearing a crescent.

" s' HENRICI DE PRENDERGEST."—*Appended to Excambion for lands in Prendergast to the Prior of Coldingham, 1275.—Dean and Chapter of Durham.*

823. PRENDERGAST, LORD HENRY.

A bend cotissed; shield in tracery. Very pretty.

" s' HENRICI DE PREDERGUST."—*Appended to an Indenture made at Coldingham, A.D. 1324.—Dean and Chapter of Durham.*

824. PRENDERGAST, JOHN.

> A fleur-de-lis ; above the shield two branches of foliage.
>
> " s' JOHIS DE PRENDIRGEST."—The two last letters are at the sides of the shield, the engraver having found the surrounding circle not large enough to contain them.
>
> The form of the fleur-de-lis here is quite a unique example of that graceful charge, the peculiarity consisting in having a sprig issuing from between the centre and dexter leaf. In very early periods it was not very uncommon to represent the flower with *two* such sprigs, as in the Montgomery Seal (No. 590, "*Descriptive Catalogue of Ancient Scottish Seals*"), and others. Whether the omission in this instance is the result of accident or design it is impossible to say.
>
> The original of this pretty seal, undoubtedly the work of the fourteenth century, is of silver, with a pyramidal shank of seven faces scalloped, and terminating in a trefoil-shaped loop, which seems rather like a later addition. It was found at Bleaton in Glenshee, ten miles north of Blairgowrie, Perthshire, during the trenching of a field, and is now in the possession of James Anderson, Esq. of Comrie Castle, proprietor of Bleaton.—*Communicated by George Seton, Esq.*

825. PRESTON, SIMON, PROVOST OF EDINBURGH.

> Three unicorns' heads erased ; in fess point a trefoil slipped, as a mark of cadency.
>
> " SYMON PRESTON." A very pretty seal.—*Appended to a Sasine dated 6th September* 1532.—*Edinburgh Charters.*

826. PRINGLE, JAMES.

> On a bend, three escallop-shells.
>
> " s' JACOBI HOPPRINGELL."—*Appended to Precept of Sasine in favour of John Bellenden, of the lands of Romano Grange, in the lordship of Newbattle, co. Peebles, 11th December* 1549.—*Principal Campbell, Aberdeen.*

827. PRINGLE, JOHN, OF SMAILHOM.

> On a bend engrailed, three escallop-shells ; foliage at the top and sides of the shield.
>
> " s' JOHANNIS HOPPRINGIL DE SMAILHAME."—*Appended to Indenture between James Heriot of Trabruae and John Pringle of Smailholm, 3d May* 1537. —*Alexander Pringle of Whytbank, Esq.*

828. PURVES, THOMAS.

On a bend, three cinquefoils; in sinister chief point a mullet; slight foliage surrounding the shield.

" S' THOME PURVAS."—*Appended to an Instrument dated* A.D. 1410.—*Dean and Chapter of Durham.*

829. QUICKSWOOD, DAVID.

A tree, or rather a bush, not on a shield.

" S' DAVI. FILI EARNOLFI DE QC. WDA."—*Appended to Charter of some lands at Aldcambus to St. Cuthbert's of Durham, circa* 1200.—*Dean and Chapter of Durham.*

There are several varieties of the seals of the Quickswoods about this period, all having the design of a bush or shrub (in no case on a shield), which may probably be considered as a punning device on the name.

830. QUINCI, ROBERT DE.

Seven mascles, three, three, and one.

" [SE]CRETUM ROBERTI DE [QUINCI]."—*Appended to " Indenture between W., the Abbot and Convent of Holyrood-house and Robert de Quincii,* A.D. 1322."—*Panmure Charters.*

831. RAIT, DAVID.

A cross invecked. The initials M · R · (Master David Rait) surround the shield.

Appended to a Reversion, 29th September 1591.—*Crawford and Balcarras Charters.*

832. RAMSAY, ALEXANDER.

An eagle displayed; foliage at the top and sides of the shield.

" S' ALEXANDRI RAMSAY."—*Appended to a Precept of Sasine,* 1593.—*Communicated by J. C. Roger, Esq.*

833. RAMSAY, JAMES. *Plate* VII. *fig.* 10.

Conché; an eagle displayed. Crest, on helmet, a unicorn's head. Supporters, two griffins segreant.

" S' JACOBI DE RAMSAY."—*Appended to a Deed by Patrick Tripnay, dated 19th June* 1401."—*From General Hutton's Collection.*

834. RANDOLPH, THOMAS.

>A device of a lion bearing a cushion on its back.
>
>" s' THOME RAN[DOLPH]."   Detached Seal.—*H.M. Record Office.*

835. RANDOLPH, THOMAS, EARL OF MORAY.

>A fine seal.   An armed knight on horseback galloping to dexter; in his right hand a sword connected with his coat of mail by a chain, on his left arm a shield bearing three cushions within the royal tressure, which charges are also repeated on the housings.
>
>" s' THOME RANULFI . . . COMITIS MORAVIA."—" *Sigillum Thomæ Ranulphi Comitis Morraviæ et Domini Vallis Anandiæ et Insulæ Manniæ Gubernatoris Regni Scotiæ, in a° 1338.*"

836. COUNTER SEAL OF THE PRECEDING.

>A shield as in the preceding.
>
>" SIGILLUM THOME COMITIS MORAVIA."—*Hutton's* " *Sigilla,*" p. 122.

837. RANDOLPH, AGNES, COUNTESS OF MORAY AND MARCH, DAUGHTER AND HEIRESS OF THE ABOVE THOMAS RANDOLPH, EARL OF MORAY, AND WIFE OF PATRICK, EARL OF MARCH,—BETTER KNOWN AS " THE BLACK AGNES."

>A fine seal, having four shields disposed in a circle, a cross between each, the base points of the shields meeting in the centre.   The first bears Scotland; the second and third seem both to have the lion rampant, for Dunbar,—her husband's; while the fourth is charged with three cushions, for Moray,—her paternal arms.
>
>" [s' AG]NETIS COETISSE. MAR[CHI]E ET MOR[AY]."—*Appended to a Document dated 24th May 1367.—Dean and Chapter of Durham.*

838. RATTRAY, ALEXANDER.

>On a chevron between three roses, one of the same inter two mullets; foliage at the top and sides of the shield.
>
>" s' ALEXANDRI RATRAY."—*Appended to a Charter dated 17th April 1628.—Mussel-burgh Charters.*

839. RATTRAY, JOHN, BAILIE OF ABERDEEN.

>A fess between three cross-crosslets fitchée, fesswise in chief, and an ermine spot (?) in base.
>
>" s' JOHANNIS RATTRA," A.D. 1504.—*David Laing, Esq.*

840. RATTRAY, SILVESTER, " DE EODEM."

Six cross-crosslets fitchée, three, two, and one.

" s' SILVESTRI DE RATTRAY."—*Appended to a Precept of Sasine of the lands of Graniche in earldom of Athol and county of Perth, in favour of John Stuart of Ferthirkile, 26th March 1465.—R. Almack, Esq., Melford, Suffolk.*

841. RENTON, JOHN.

A chevron between three buckles (?).

" s' JOANNES RENTON."—*Appended to Inquisition of the lands of Aldengray on behalf of William Paxton, A.D. 1429.—Dean and Chapter of Durham.*

842. RENTON, JOHN.

A lion rampant. Foliage at the top and sides of the shield.

" s' JOHANNIS DE RENTON."—*Appended to Perambulation of the lands of Adam Forman, A.D. 1430.—Dean and Chapter of Durham.*

RESTALRIG. *Vide* LESTALRIG.

843. RICHARDSON, PATRICK, OF DRUMSHEUGH.

Two arrows in saltire, points in chief.

" PATRICIUS RICHARTSON."—*Appended to a Charter dated 9th June 1507.—Edinburgh Charters.*

844. RICHARDSON, ROBERT, COMMENDATOR OF ST. MARY'S ISLE.

A saltire, with a boar's head erased in the chief canton, and a galley (lymphad) in the base.

" s' M. ROBERTI RICHARTSOUN."—*Appended to a Charter to the Burgh of Musselburgh, 26th March 1566.—Musselburgh Charters.*

845. RICHARDSON, SIR JAMES, OF SMEATON.

A saltire, with a bull's head erased in the chief canton, and a crescent in the base. Thistle and rose ornament surrounds the shield.

" s' JACOBI RICHARDSONE DE SMET."—*Appended to a Charter by Sir James Richardson to the Burgh of Musselburgh, A.D. 1627.—Musselburgh Charters.*

846. RIND, ALEXANDER.

Ermine ; a cross indented or engrailed ; in the sinister chief a mullet.

" s' ALEXANDRI RIND."—*Appended to the Charter of foundation of an altar to St Salvator in the Church of St. Giles, Edinburgh, 29th August 1512.—Edinburgh Charters.*

847. ROBERTSON, JOHN, Treasurer of Ross.

A chevron between three wolves' heads erased.

" s' m. johannis robartson."—*Appended to an Instrument regarding the Teinds of Ferenton, A.D.* 1589.—*Crawford and Balcarras Charters.*

848. ROCASTEL, WILLIAM.

Oval-shaped seal. Within a Gothic niche, a monk kneeling before a saint, holding in the right hand a palm branch or sword, and in the left a scroll, with inscription now illegible; in base an eagle with a bird in its beak.

" s' willi. de rocastle . . . ."—*Appended to the same Homage as No.* 48, A.D. 1292.—*H.M. Record Office.*

849. ROGER (?), SIR WILLIAM.

Broken. Couché; a stag's head erased, with a mullet in front of its mouth. Above the shield a helmet and crest; the latter is nearly lost, but it has probably been a stag's head erased. Supporters, two lions sejant, gardant; the background seems to have been ornamented with foliage.

" s' wilelmi . . . ."—*From a cast in the collection of the late Charles Roger, Esq., of Dundee, who gives the date* " 1478."—*Communicated by J. C. Roger, Esq.*

850. ROGER, SIR WILLIAM, same person as the last (?).

Shield and supporters as in the preceding; crest, on helmet, a stag's head erased, pierced through the neck with an arrow, point to dexter.

" s' . . . . roger."—*From a cast in the collection of the late Charles Roger, Esq., Dundee, who gives the following note:*—" *Sir Wm. Roger, Knt., Privic Councellour to James jii., King of Scots,* 1479," . . . " *from a Chartour of Renounciation be him in favours of his sone Wilm. (thoiraftir Sir William), be his spouse Joneta Valence,* A.D. 1479."—*Communicated by J. C. Roger, Esq.*

851. ROGER, SIR WILLIAM, Knight, Son of the above Sir William Roger (?).

A shield only, bearing the same charge as in the two preceding.

" s' w . . . roger, mil."—*From a cast in the collection of the late Charles Roger, Esq., Dundee, who gives the note,* " *Sir William Roger, Kt., from an Instrument dated* 1533, *concerning or conveying a piece of ground within the parish of Galstoun.*"—*Communicated by J. C. Roger, Esq.*

852. ROLLOK, HERCULES, Burgess of Dundee, Master of the High School of Edinburgh, A.D. 1584; died 1599.

A chevron between three boars' heads and necks erased; foliage at the top and sides of the shield. Very rudely executed.

" s' m. hercull rollok."—*Appended to Procuratory of Resignation of a tenement in Dundee in favour of John Schewan, merchant burgess of Dundee, 13th November 1593.—J. C. Roger, Esq.*

853. ROMANOES, JANET.

Three boars' heads and necks erased.

" s' joneta romanos, of yat ilk."—*Appended to a " Tack" of Cathope, in Peebles, a.d. 1515.—Blackbarony Charters.*

854. ROSS, HUGH, of Rariches, second Son of Hugh Earl of Ross.

Three lions rampant within a bordure charged with eleven escallop-shells (ermine spots ?), and in base point a mullet.

" s' hugonis de rosse."—*Appended to Charter of the lands of Scatterly and Bach, in Buchan, to Peter Graham, a.d. 1351.—Seaforth Charters.*

855. ROSS, JOHN.

Very much injured, but has evidently been a fine design. Three lions rampant, apparently contourné; above the shield two eagles pluming themselves. Inscription illegible, a.d. 1431.—*Kilravock Charters.*

856. ROSS, JOHN, of Auchlossin.

Two water-bougets in chief.

" s' joh. le rosi de achlossin."—*Appended to Procuratory by John le Ross of Auchlossin, for resigning all lands belonging to him in favour of William, Thane of Cawdor; not dated, but probably of same date as the Charter, 1457.—Cawdor Charters.*

857. ROSS, JOHN, Burgess of Linlithgow.

On a chevron, between three water-bougets, a rose ; in middle chief point a crescent.

" s' johannis ros."—*Appended to Sasine of Lochmylne by James Cockburn of Langton to William Denniston, burgess of Linlithgow, a.d. 1540.—Communicated by J. T. Gibson-Craig, Esq.*

858. ROSS, ROBERT, of Craigie.

A fess chequé between three water-bougets.

" s' robertus rose."—*Appended to Letter of Robert Ross, " domini de Cragy," to the Prior of Coldingham, regarding lands at Paxton and Halynecrayne. " Apud Perthe, 1423."—Dean and Chapter of Durham.*

859. ROSS, WILLIAM, EARL OF.
Three lions rampant.
" s' WILLELMI COMITIS DE ROSS."

860. COUNTER SEAL of the preceding.
An antique gem. A man and horse, rudely executed.
" s' DOMINI WILELMI COMITIS DE ROSS." Detached Seal.—*H.M. Record Office.*

861. RUTHERFORD (?), BINER.
An eagle, wings expanded, not on a shield.
" s' BINERI DE TOCHET FORD (?)."—*Appended to the same Deed of Homage as No.* 48, A.D. 1296.—*H.M. Record Office.*

862. RUTHERFORD, DAVID.
On a fess, between a mullet in chief and a boar's head and neck erased in base, three birds (martlets ?).
" s' DAVID RUTHIRFURD."—*Appended to Charter by James Learmonth, Chaplain of the Altar of the Blessed Virgin Mary at St. Andrews, and Elizabeth Balfour his spouse, with consent of the Council of St. Andrews, A.D. 1554.—St. Andrew's Charters.*

863. RUTHERFORD, WILLIAM.
A bull's head cabossed, between the horns a man's head affronté, not on a shield.
" s' WILLMI ROTHIRFORD." Detached Seal.—*H.M. Record Office.*

864. RUTHVEN, WILLIAM, PROVOST OF PERTH.
Paly of six, and a label of four points.
" s' VILELMI RUTHVEN."—*Appended to Sasine of the earldom of Menteith in favour of Margaret, daughter of Henry VII., as her proper marriage gift,* 29th May 1503.—*H.M. Record Office.*

865. RUTHVEN, WILLIAM.
Paly of six, or three pales. Crest, a stag's head.
" s' WILELMI DE RUTHVEN."—*Appended to Ratification of Truce for one year, between King Henry VIII. and King James V.,* 7th October 1518.—*H.M. Record Office.*

866. RUTHVEN, WILLIAM, LORD.
Three pallets, the centre one charged with a mullet.
" s' WILLIELMI DOM. RUTHVEN."—*Appended to Charter by William Ruthven of Lounan, etc. etc.,* 27th June 1527.—*Hutton's "Sigilla,"* p. 120.

867. RUTHVEN, WILLIAM, of Lunan.

Three pallets, the centre one charged with a human heart.

" s' WILLIELMI RUTHVEN DE LUNAN."—*Appended to the same Document as the preceding.*

868. RUTHVEN, WILLIAM, second Lord Ruthven, married to Janet Haliburton, Lady of Dirleton. *Plate V. fig. 9.*

Couché ; three pallets. Crest, on a helmet, a ram's head issuing from a coronet. Supporters, two goats.

" SIGILLUM WILELMI DNI DE RUTHIE D."—*Appended to Precept of Clare Constat by Janet Lady Dirleton, with consent of her husband, William Lord Ruthven, to Elizabeth Scrymgeour, 17th May 1547.—Wedderburn and Birkhill Charters.*

869. RUTHVEN, WILLIAM, fourth LORD, Lord High Treasurer of Scotland, created Earl of Gowrie 1581.

Couché ; three pallets (probably paly of six). Crest, on helmet with mantlings, a ram's head. Supporters, two rams ; below the shield the motto, " DEID CHAU."

" s' VILELMUS DNS RUTHVEN ET DIRLTOUN."—*Appended to Precept of Sasine, 20th November 1580.—David Laing, Esq.*

870. ST. MICHAEL, JOHN.

Oval-shaped. In the upper part, St. Michael vanquishing the dragon ; in the lower, a monk, with a cross in his hand, seems to arrest the progress of a man on horseback, with a hawk on his left hand.

"— s' JOHANNIS DE SCO. MICHAELE."—*Appended to Homage Deed, 20th July 1295.— H.M. Record Office.*

871. SANDILANDS, JAMES, Lord Torphichen.

Much-damaged impression, but has evidently been a fine seal. Per fess, in chief a crown, in base a thistle, being the coat of augmentation assigned to Sir James Sandilands as Grand Prior of the Knights of St. John of Jerusalem (Knights of Malta), and which is now usually carried in the first quarter of the family shield, thus taking precedence of the ancient coat of Sandilands. Crest, on full-faced helmet, an eagle displayed. The shield is supported by a single lion sejant on the dexter side.

" s' JACOBI SANDILANDIS DOMINE DE TORPHICHINE."—*Appended to Precept of Sasine of the lands of Franishill, co. Peebles, 28th May 1569.— Blackbarony Charters.*

872. SANDILANDS, SIR JAMES, of Slamannan, Knight.

> Quarterly: first and fourth, a bend, for Sandilands; second and third, a human heart, and on a chief three mullets, for Douglas; foliage around the shield.
>
> " s' D. JACOBI SANDELANDIS DE SLM."—*Appended to Charter by Sir James Sandilands, Knight, of the lands of Corswoodhill, etc., in barony of Calder and sheriffdom of Edinburgh, to Robert Williamson, 20th October 1598.—Stewart's Hospital Charters.*

873. SANDILANDS, JAMES.

> Quarterly: first and fourth, a human heart, and on a chief three mullets, for Douglas; second and third, a bend (erroneously placed sinisterwise) for Sandilands.—*Appended to an Instrument of Sasine dated A.D. 1524.—St. Leonard's College Charters.*

874. SANDILANDS, PETER.

> Quarterly, as in the last.
>
> " s' M. PETRI SANDELANDIS."—*Appended to " an Instrument by Peter Sandilands, Canon of the Cathedral Church of Ross, dated 25 Jany. 1543, in the Archives of the Burgh of Peebles."—Hutton's " Seals," p. 94.*

875. SCHEVES, JOHN, " Doctor in Degree."

> An antique gem—an Egyptian scarabæus.—*Appended to " Concordia" by the Commissioners for Governing the Marches, 12th July 1429.—H.M. Record Office.*

876. SCHEVES, JOHN.

> Quarterly: first and fourth, three cats-a-mountain passant in pale, for Scheves; second and third, a cross.
>
> " SIGILL. JOHANNIS SCEVES."—*Appended to the same Instrument as No. 740.*

877. SCOTT, SIR WILLIAM, of Balwerie, Knight.

> A chevron between three lions' heads erased. The shield in centre of tracery.
>
> " s' VILELMI SCOT MILITIS."—*Appended to Reversion of West Raith, in Fife, to Sir David Weemys, Knight, 12th November 1509.—Communicated by W. Fraser, Esq.*

878. SCOTT, SIR WILLIAM, of Balwerie, Knight.

Conché; three lions' heads erased. Crest, on helmet, a hand grasping a sword.

"S' VILLI SCOTTE DE BALVERY MILITIS."—*Appended to Truce between the Ambassadors of King James V. and King Henry VIII. 4th September* 1524.—*H.M. Record Office.*

879. SCOTT, SIR WALTER, of Branxholm and Kirkurd, Knight.

A bend charged with two crescents, and in the upper part a mullet.

"S' VA[LTE]RI SCOT MILES."—*Appended to an Instrument dated* A.D. 1529.—*Buccleuch Charters.*

880. SCOTT, SIR WALTER, of Buccleuch, created Earl of Buccleuch 16th March 1619.

A bend charged with two crescents, and a mullet in the upper part. A rose and thistle ornament surrounds the shield.

"S' VALTERI D' SCOT DE BUKCLUGH."—*Appended to Confirmation by Sir Walter Lord Scott of Buccleuch, as Superior, of grant of liferent annuity, by William Baillie of Cairnbrow, to Margaret Jackson his wife, out of the lands of Cairnbow, dated* 21st *January* 1619.—*Principal Campbell, Aberdeen.*

881. SCOTT, WALTER, the same person as the preceding. *Plate* VI. *fig.* 6.

A bend charged with two crescents, and in the upper part a mullet; above the shield a coronet; supporters, two ladies vested.

"S' VALTERI CO. DE BUKCLEUGH DO. SCOT QUHITCHESTER ET YKSDAIL."—*Appended to Precept of Clare Constat, 27th October* 1632.—*Buccleuch Charters.*

882. SCOTT, FRANCIS, second Earl of Buccleuch, Son of the first Earl. *Plate* VI. *fig.* 7.

On a bend, a mullet of six points between two crescents; above the shield an earl's coronet. Crest, on helmet with mantlings, a stag trippant, and in front of it the word "AMO." Supporters as in the last; the background filled up with foliage.

"S' FRAN. COM. DE BUCCLECCHE DOM. SCOTT DE VICHESTER ESKDALE ET DOM. REGAL. DE DALKEITH," A.D. 1648.—*From the original brass matrix, which was found a few years since among a lot of old metal sold to a dealer in Stirling, and is now in the possession of the Duke of Buccleuch.*

T

883. SCOTT, FITZROY, JAMES, Duke of Monmouth, Natural Son of King Charles II., married to Lady Anne Scott, second Daughter and heiress of the above Earl Francis, whose Surname of Scott he assumed. *Plate VI. fig. 8.*

A fine seal, though rather injured. Quarterly: first and fourth, Scotland; second quarterly, France and England; third, Ireland; over all a baton sinister. On an escutcheon surtout the arms of Scott, as before. The shield is surrounded by the Garter, with the motto of the Order, and surmounted with a ducal coronet. Crest, on a helmet with mantling, a lion statant, gardant, crowned, on a ducal coronet. Supporters, dexter, a unicorn gorged with a coronet, and chained; sinister, a stag, as in the dexter.

"SIGILL. JACOBI DUCIS DE BUCCLEUGH ET DE MONMOUTH."—*Appended to a Precept of Clare Constat in favour of Margaret and Agnes Matherstoun, 4th January 1669.—Communicated by W. Fraser, Esq.*

884. SCOTT, WILLIAM, of Harden.

On a bend, a crescent between two mullets, with slight foliage surrounding the shield.

"s' WILLLMI SCHOTE."—*Appended to Wadset of the lands of Nether Harden, 21st October 1540.—Buccleuch Charters.*

885. SCOTT, WILLIAM, Constable-Depute of Montrose.

A chevron between two wolves' heads in chief, and a cinquefoil in base.

"SIGILLUM WILMI SCOT."—*Appended to a Sasine dated 25th April 1513.—Brechin Charters.*

886. SCOTT, JOHN, Burgess of Edinburgh.

On a bend, a mullet between two crescents; in the dexter base, a mullet.

"s' JOHANNIS SCOTT."—*Appended to a Charter by John Scott of a tenement, etc., on the north side of the Castle Hill, 26th March 1597.—Edinburgh Charters.*

887. SCRYMGOUR, JAMES, of Dudhope, Constable of Dundee.

Couché; a lion rampant, holding in its dexter paw a sword. Crest, on helmet with mantling, a hand (lion's paw ?), holding a sword sinister bendwise.

"s' JACOBI [SCRYMGOUR]."—*Appended to a Sasine of a tenement in Dundee. A.D. 1578.—Communicated by J. C. Roger, Esq.*

888. SCRYMGOUR, SIR JOHN, of Dudhope, Constable of Dundee, Son of the above James Scrymgour.

Precisely the same design as the last.

" s' JOHANNIS SKRIMGEUR."—*Appended to a Charter by Robert Jack, merchant burgess of Dundee, son of Thomas Jack, to John Bathgait, Notary-Public, Dundee, and Janet Jack his spouse, of a tenement in Dundee, 3d November 1632.—Communicated by J. C. Roger, Esq.*

889. SEMPILL, WILLIAM, SECOND LORD. *Plate V., fig. 3.*

Couché; a chevron counter compony between three hunting-horns stringed. Crest. on helmet, a stag's head; supporters, two greyhounds.

" s' WILELMI DNI SEMPIL."—*Appended to the same Instrument as No. 723.*

890. SETON, SIR ALEXANDER, OF SETON, KNIGHT.

An armed knight on horseback, galloping to the dexter, with a sword in his right hand, and on his left arm a shield, charged with three crescents.

" SIGILL. ALEXANDRI DE SETON."—*Appended to " Carta Domini Alexandri Settone de eodem militis facta Adæ de Polibeny terrarum de Beth in liberam maritagium cum Emma sorore mea," c. 1250.—Hutton's " Sigilla," p. 135.*

891. SETON, SIR ALEXANDER, OF SETON, KNIGHT.

Three crescents within the royal tressure. The top and sides of the shield ornamented with foliage.

" SIGILLUM ALEXANDER [DE SETON]."—*Appended to Charter by Sir Alexander Seton to the Church of St. Mary's, Haddington, of twenty shillings annually from the Mill of Barins, A.D. 1337.—" Harl. 4693."—Hutton's " Sigilla," p. 110.*

892. SETON, WILLIAM, LORD.

Seton, as before. The shield in centre of tracery.

" SIGILL. WIL. DNE. SETON."—*Appended to " Carta Willielmi Domini de Settone confir. unius tenementum in Langue Nidryffe Do. Patricio Grey," etc., 6th January 1381.—Hutton's " Sigilla," p. 173.*

893. SETON, ROBERT, LORD, CREATED EARL OF WINTON, 16TH NOVEMBER 1600.

Quarterly; first and fourth, three crescents, for Seton; second and third, three garbs, for Buchan, all within the royal tressure.

" s' ROBERTI DOMINI DE SETOUN."—*Appended to a Charter by Robert Lord Seton of a house, etc., at Tranent, to Andrew Nicolson and Janet Gibson his spouse, 12th February 1600.—Edinburgh Charters.*

894. SETON, GEORGE, third Earl of Winton, second Son of the preceding. *Plate* I.,
    Frontispiece, fig. 6.

> Quarterly : Seton and Buchan, as before. On an escutcheon surtout, a star of
> twelve points within the royal tressure, for the title of Winton. Crest, on an
> earl's coronet, above a full-faced helmet, with mantling, a dragon spouting fire,
> wings elevated, and charged with a mullet. Supporters, two foxes collared and
> chained. Inscription, on a scroll above the crest, "s' GEOR. CO. DE WINTOUN D.
> SETOUN," and on a ribbon below the shield the motto, " HAZARD ZIT FORDWARD."
> —*Appended to a Charter of Aldenton, 21st June* 1608.—*Communicated by
> W. Fraser, Esq.*

895. SETON, ALEXANDER, of Tullibody, and Lord of Geddes.

> Quarterly : first and fourth, three garbs ; second and third, three crescents.
> " s' ALEXANDRI CITON." Very rudely executed —*Appended to Charter by Alexander
> Seton to William, Thane of Cawdor, of the lands of Meikle Geddes, etc., in the
> earldom of Moray and sheriffship of Nairn, 24th October* 1493.—*Cawdor Charters.*

896. SETON, JOHN, first Baron of Cariston, co. Fife,
    second Son of George, sixth Lord Seton.

> Three crescents within the royal tressure. The crescent
> in the dexter chief charged with a bezant as a differ-
> ence. (*See* Nisbet's *Essay on Armories,* p. 108.)
> Crest, on helmet with mantling, a dragon spouting
> fire. Motto above the crest, " HAZARD ZIT FORD-
> WARD," *circa* 1552.—*From a steel seal in the posses-
> sion of George Seton, Esq., Advocate, representative
> of the family of Cariston.*

897. SETON, ALEXANDER, Lord President of the Court of Session, and Lord
    High Chancellor of Scotland, created Earl of Dunfermline, A.D. 1605.
    Plate V. *fig.* 5.

> Quarterly : first and fourth, Seton, as before ; second and third, on a fess, three
> cinquefoils, for Hamilton. (The Chancellor's mother was Isabella, daughter
> of Sir William Hamilton of Sanquhar, High Treasurer of Scotland.) Above
> the shield an earl's coronet. Crest, on helmet with mantlings, a crescent, and
> on a ribbon the word " SEMPER." Supporters, two horses at liberty. In the
> background of the lower part of the seal is a view of the city of Dunfermline.
> " SIGILLUM ALEXANDRI SETONII FERMELINODUNI COMITIS," ETC.—*This fine seal is
> appended to a Charter to James Earl of Abercorn, A.D.* 1618.—*Communicated
> by W. Fraser, Esq.*

898. SETON, ALEXANDER, THE SAME PERSON AS THE PRECEDING.

A shield only. Quarterly: Seton and Hamilton, as before, but in this instance the coat of Seton occupies the second and third quarters. Foliage at the top and sides of the shield.

" S' ALEX. COMITIS DE DUNFARM."—*Appended to Charter by Alexander Earl of Dunfermline to Sir John Campbell of Cawdor, of the Kirk lands of Durris, in the lordship of Urquhart, co. Inverness, 12th February 1610.—Cawdor Charters.*

899. SETON, ALEXANDER, THE SAME PERSON.

An uncommonly pretty seal, or small signet. A cinquefoil within a crescent, being a combination of the principal charges of his arms, under a coronet. The same design occurs, in relief, within a shield, on a stone close to the outside of the south door, on the east, of the Abbey Church, Dunfermline, with the date 1607.—*From a Letter to the King (James VI.), dated 8th January 1605.—Sir James Balfour's Collection, Advocates' Library.*

900. SETON, SIR WILLIAM.

On a chevron, between three crescents, a rose or cinquefoil within the royal tressure; above the shield, in place of the crest, the initials $_{W.}^{S}._{S.}$ (Sir William Seton). —*From a Letter to Viscount Annan, dated 12th August 1623.—Sir James Balfour's Collection, Advocates' Library.*

901. SETON, WILLIAM, OF HIS MAJESTY'S GUARD OF HORSE, DESCENDED FROM THE FAMILY OF SETON OF MELDRUM.

A lance, bendwise, point imbrued, between three crescents, all within the royal tressure. Crest, on helmet with mantlings, an arm, in armour embowed, the hand grasping a spear. Motto, " MAJORUM VESTIGIA PREMO."—*From a well-executed three-sided, revolving steel seal, in the possession of George Seton, Esq., Advocate.*

The two other sides are respectively occupied by the crest and motto as above, and the initials W. S., surmounted by two Cupids bearing a wreath of flowers, very prettily designed.

902. SETON, OF TOUCH.

Quarterly: first and fourth, Seton, as before; second and third, three escutcheons for Hay. Crest, on a helmet with mantling a boar's head and neck erased. Supporters, two greyhounds. Motto, above the crest, " FORWARDS OURS.—*From a steel seal in the Museum of the Society of Antiquaries of Scotland.*

903. SHAW, WILLIAM, Provost of the Collegiate Church of Abernethy.

Three covered cups, of very peculiar form; in chief a mullet; foliage at the top of the shield.

"s' vilelmi schau p. de abirneti."—*Appended to the same Instrument as No.* 411.
—*David Laing, Esq.*

904. SIME, ADAM, Chaplain of St. Ninians, Leith.

Two wolves' (?) heads erased, contourné; a mullet in chief, probably as a difference only; in base a chalice; foliage at the top and sides of the shield.

"s' domeni adame sime." Rudely executed.—*Appended to a Sasine, a.d.* 1545.—
In the Museum of the Society of Antiquaries of Scotland.

905. SIM (or SYME), WILLIAM, one of the Bailies of Edinburgh.

A chevron between two mullets of six points in chief, and an axe (Lochaber?) in base.

"s' vilelmi sym b."—*Appended to a Charter by Robert Haitheway of all his right of property in the city of Edinburgh to William Lauder, burgess of said burgh, and Janet Grey his spouse, 27th October* 1527.—*Edinburgh Charters.*

906. SIMPSON, GEORGE.

A fess between three crescents.

"s' georgi symson."—*Appended to Charter of Quarter of East Loch, Peeblesshire, to William Gibson, 13th April* 1561.—*Blackbarony Charters.*

907. SINCLAIR, AGNES, Daughter of Henry Lord Sinclair, Wife of Patrick, third Earl of Bothwell, and Mother of James Earl of Bothwell, Husband of Queen Mary. (The Countess of Bothwell was usually designated as "Lady of Morham." A divorce took place in 1543, but she retained the title of Countess of Bothwell till her death in 1572.)

Quarterly: first and fourth, a galley within a double tressure, flowered and counter flowered; second and third, a galley in full sail. On an escutcheon surtout, a cross engrailed, for Sinclair. A scroll ornament at the top of the shield, and at each side a griffin.

"s' agneti sinkler dna de morha."—*Appended to Precept of Sasine, a.d.* 1564.—
Smeaton Charters.

908. SINCLAIR, JOHN, Dean of Restalrig, afterwards Lord President of the Court of Session; Bishop of Brechin, a.d. 1563; died 1566.

A cross engrailed.

"s' m. johannes sinkler," a.d. 1543.—*Elibank Charters.*

909. SKIRVING, DUNCAN.

A sword fesswise, point to dexter between three boars' heads and necks erased; foliage at the top and sides of the shield.

" s' DUNCANI SKRYVIN."—*Appended to Reversion by Mr. James Fullerton, parson of Edzell, and Isabella Graham his spouse, to Sir D. Lindsay of Edzell, of an annual rent of fifty merks forth of the lands of Meikle Mengry, 10th November 1589.—Crawford and Balcarras Charters.*

910. SMALL, MARIOTTE.

A saltire between a mullet in chief and a crescent in base.

" MARIOTE SMAL."—*Appended to an Instrument dated 13th August 1584.—Edinburgh Charters.*

911. SOLET, ROBERT.

Merely a scroll ornament, not on a shield.

" s' ROBERTI SOLET."—*Appended to the same Homage as No. 48, A.D. 1296.—H.M. Record Office.*

912. SORASBURA, JOHIS DE.

A wheel on flower ornament.

" s' JOHIS D. SORASBURA."—*Appended to the same Instrument as the last.*

913. SPALDING, DAVID.

On a cross, a cross-crosslet, fitchée.

" s' DAVID DE SPALDING."—*Appended to the Endowment of a Chaplaincy for the Altar of St. Margaret in the Church of the Virgin Mary, Dundee, by David Spalding, jun., burgess of Dundee, 13th January 1445.—Communicated by Lord Lindsay.*

914. SPENCE, DAVID (ONE OF THE BAILIES INFEFTING).

A lion rampant, debruised with a bend charged with a buckle between two mullets, the shield between the letters D · S ·—*Appended to Sasine of Netherton of Dunfermline, 16th June 1549.—Elgin Charters.*

915. SPENCE, HUGH.

Fretty, and on a chief dancette, three roses.

" s' HUGONIS DE SPENS."—*Appended to an Inquisition of the Lands of Alexander Lumsden, as heir to his brother, Thomas Lumsden, A.D. 1444.—Dean and Chapter of Durham.*

916. SPENCE, JOHN, of Condy, Advocate to Mary Queen of Scots.

> On a chevron, between three boars' heads and necks erased, a thistle stemmed and leaved. On a scroll above the shield is a motto, the only letters remaining are c · t · a ·
>
> " ·' mori joannis s[pen]s de condy."—*Appended to the same Instrument as No. 682.*

917. STIRLING, LUKE, of Botchquhumore, first of Keir.

> On a bend, three buckles, tongues pointing upwards.
>
> " sigillum lucas de strivelin."—*Appended to Procuratory of Resignation of the lands of Byazhartg, etc., in co. Fife, to George Leslie, Lord Leven, 6th May 1448.—Rothes Charters.*

918. STIRLING, SIR JOHN, of Keir, Knight.

> On a bend, three buckles, tongues pointing downwards.
>
> " s' johannis stervelin mil."—*Appended to various Instruments between a.d. 1520 and 1540.—Keir Charters.*

919. STIRLING, SIR JAMES, of Keir, Knight, Son of the above Sir John Stirling.

> On a bend, three buckles, tongues pointing downwards. Foliage at the top and sides of the shield.
>
> " s' jacobi strenvilang."—*Appended to various Instruments between a.d. 1540 and 1550.—Keir Charters.*

917.   918.   919.

920. STIRLING, SIR JAMES, of Keir, Knight. Same person as the last.

> Very much damaged, but evidently the same blazon as in the preceding examples.
>
> " s' jacobi strevelinge de keir miltis."—*Appended to Procuratory of Resignation of the lands of Cadder, 1579.—Keir Charters.*

921. STIRLING, SIR JAMES, OF KEIR, KNIGHT. *Plate V. fig. 8.*

A third seal of the same person, and precisely the same charges as those in the preceding seals, but in this the tongues of the buckles are pointing upwards, the shield of a different shape, and above it an elaborate foliage.

" S' JACOBI STRIVILING MILITIS DE KEIR."—*Appended to a Charter by Sir James Stirling of Keir to Archibald Stirling, his second son, of the barony of Keir, sheriffdom of Perth, 15th September 1579.—Keir Charters.*

922. STIRLING, SIR ARCHIBALD, OF KEIR, KNIGHT, SECOND SON OF THE ABOVE SIR JAMES STIRLING.

On a bend engrailed, three buckles, tongues pointing upwards. A rose and thistle ornament at the top and sides of the shield. The engrailing of the bend in this example is doubtless as a mark of cadency, and has continued to be carried by the family of Keir till the present time."

" S' ARCHIBALDI STIRLYNG DE KEIR."—*Appended to various Instruments between A.D. 1589 and 1639.—Keir Charters.*

923. STIRLING, WILLIAM, OF CADDER.

Couché; on a bend engrailed, three buckles. Crest, on helmet with mantlings, a swan's neck issuing from a coronet.

" S' WILLELMI STRIVELIN [DE CAD]DER."—*Appended to a Procuratory of Resignation by William Stirling of the lands of Kirkmichael and Blarnara in favour of William Stirling, his son and heir apparent, 7th January 1492.—Keir Charters.*

924. STIRLING, GEORGE, YOUNGER OF CRAIGBERNARD OR CRAIGBERNET.

On a bend, between two saltires couped, a mullet inter two buckles, tongues pointing upwards; in middle, chief point a rose as a difference.

" S' GEORGII STIRLING."—*Appended to Procuratory of Resignation of the lands of Kilbrynet, with the mill, in the lordship of Lennox and sheriffdom of Stirling, to Matthew Earl of Lennox, 16th March 1502.—Keir Charters.*

U

925. STIRLING, JOHN, Younger of Tullidovy.
> On a bend sinister, between two mullets, three of
> the same.
>> "S' JHONE STEIRVELING."—*Appended to Charter of
>> some lands in the barony of Reskoby, in Forfar-
>> shire, by John Stirling, younger, with consent of
>> Katherine Lindsay his spouse, to Alexander
>> Rattray, 23d October 1539.—Communicated by
>> W. Fraser, Esq.*

926. STRACHAN, JAMES, Canon of the Chapter of Aberdeen.
> A chief bearing three cinquefoils; foliage at the top and sides of the shield.
>> "S' M. JACOBI STRATYAUTHY."—*Appended to an Instrument concerning the parson-
>> age of Memuuir, A.D. 1565.— Crawford and Balcarras Charters.*

927. STRACHAN, JOHN.
> A stag courant to sinister between three roses, not on a shield.
>> "SIGILL. JOHANIS DE STRAUCHIN."—"*Carta donationis Johannis de Strauchine facta
>> Domino Alexandro de Settone de eodem Milite de Vitali servitio suo, contra
>> omnes mortales Dom. Regem et heredes, suos exceptis, etc.—Apud Perth, 7th
>> August, in a° 1360.—Sigillum meum est appensum."—Hutton's "Sigilla,"
>> p. 170.*

928. STUART, ALEXANDER.
> A fess chequé.
>> "S' ALEXSANDRE [DE SENESCALL]"—*Appended to Homage Deed, A.D. 1292.—H.M.
>> Record Office.*

929. STUART, JOHN, Earl of Angus. *Plate I. Frontispiece, fig. 3.*
> A fine seal, but much injured. Two winged lions supporting a shield bearing a fess
> chequé, surmounted by a bend charged with three buckles; rich tracery sur-
> rounds the design.
>> "S' JOHANNES S ENESCALLI COMITIS ANGUSIE."—*Appended to an Instrument on the
>> Ransom of David II., 5th October 1357.—H.M. Record Office.*

930. STUART, WALTER, Earl of Athol, Son of King Robert II.
> Quarterly: first and fourth, Scotland, with a label of three points; second and
> third, paly of six, for Athol; foliage in the background. Inscription not read.
> —*Appended to Letter of the Earl of Athol, stating that his son David is about
> to proceed to England as one of the hostages for the ransom of King James,
> 28th March 1424.—H.M. Record Office.*

931. STUART, THOMAS, NATURAL SON OF KING ROBERT II., ARCHDEACON OF ST. ANDREWS.

A fine example. Scotland, surmounted with a bend counter compony (as mark of illegitimacy ?); an angel supporting the shield from above, and a dragon sejant at each side.

"S' THOME SENESCALLI [ARCHI DEA] SANCTI ADREE."—Appended to Charter by James, Prior of the cathedral church of St. Andrews, with consent of the Chapter, to Janet Douglas, of a tenement in St. Andrews, A.D. 1443.—St. Salvator's College Charters.

932. STUART, SIR WALTER, LORD OF INNERMETH, AND BARON OF THE BARONY OF ENVERKELOR, KNIGHT.

Quarterly: first and fourth, a buckle, the tongue fesswise, indicative of the descent from the family of Bonkil; second and third, a fess chequé of four tracts, for Stuart.

"S' WALTERI STUART."—Appended to Charter by Sir W. Stuart, of half of the lands of Rynde and Perisfield, in the barony of Enverkelor and shire of Forfar, to Walter Ogilvy, natural son of Margaret Paniter of Rynde, 24th April 1484.—H.M. General Register House.

933. STUART, ANDREW, FIRST LORD AVONDALE, SON OF SIR JOHN STUART, AND GRANDSON OF MURDOCH DUKE OF ALBANY.

Couché. Quarterly: first, Scotland; second, a fess chequé, and a label of four points, for Stuart; third, a saltire, cantoned with four roses, for Lennox; fourth, a lion rampant, for Fife.

"S' ANDREE STUART DNI AVENDALE."—Appended to Obligation by four marquises of Scotland, on the part of the King (James III.), for security and indemnity of the Duke of Albany, 2d August 1482.—H.M. Record Office.

934. STUART, HENRY, SON OF ANDREW, THIRD LORD AVONDALE, CREATED LORD METHVEN, A.D. 1528, ON HIS MARRIAGE WITH QUEEN MARGARET, WIDOW OF JAMES IV. Plate VI. fig. 5.

Couché. Quarterly: first and fourth, Scotland; second, a fess chequé, for Stuart; third, Lennox, as before; on an escutcheon surtout, a lion rampant, holding a castle in its paws, for title of Methven (?). Above the shield, a helmet with mantlings and a wreath, but no crest.

"S' HENRICI DNI DE METHVEN."—*Appended to Charter of the lands of Balquhidder, Perthshire, to Janet Stuart, Countess of Argyle, 16th October 1551.—Argyle Charters.*

935. STEWART, JAMES, NATURAL SON OF KING JAMES V., CREATED EARL OF MURRAY, 1562; APPOINTED REGENT OF SCOTLAND 1567; KILLED AT LINLITHGOW, 1569-70.

The arms of Scotland between the initials, I · S.—*From a Letter to Lord Robert Dudley, Master of the Horse to the Queen's Majesty of England, dated 26th February 1561.—David Laing, Esq.*

936. STEWART, FRANCIS, PROBABLY THE SAME PERSON AS THE FOLLOWING.

The arms of Scotland debruised with a ribbon, or rather a baton dexter, doubtless as mark of illegitimacy, though placed in the wrong direction; the shield surrounded with foliage.

"SIGILLUM FRANCISCI DNI DE BADZENACH ET EYNZE."—*Appended to Charter of lands of Rathnakine, A.D. 1565.—Fleurs Charters.*

The original brass matrix of this seal is in Dr. Rawlinson's Collection, Bodleian Library, Oxon.

937. STEWART, FRANCIS, CREATED EARL BOTHWELL 1587 (SON OF JOHN, PRIOR OF COLDINGHAM, WHO WAS A NATURAL SON OF KING JAMES V.)

Quarterly: first and fourth, a bend, for Vaus; second and third, Hepburn; on an escutcheon surtout, Scotland. The initials F E · B· surround the shield. Detached Seal.—*H.M. Record Office.*

938. STUART, LUDOVIC, SECOND DUKE OF LENNOX, GREAT CHAMBERLAIN OF SCOTLAND. *Plate VIII. fig. 2.*

A fine large seal. Quarterly: first and fourth, three fleurs-de-lis, within a bordure engrailed, for D'Aubigny; second and third, a fess chequé within a bordure charged with eight buckles, for Stuart of Darnley; on an escutcheon surtout, a saltire engrailed, cantoned with four roses, for Lennox. Crest, on a full-faced helmet, a bull's head breathing out flames. Supporters, two wolves rampant. The shield, etc., placed within a cloak, lined with ermine. On a ribbon issuing from the coronet is "AVANT DARNLY."

"S' LUDOVICI DUS LENOX CO. DARN. DO. TARBOL. METH. & OBING. MA. CAMER. SCOTIE." —*Appended to Confirmation of the lands of Letter, in Stirling, to Sir James Stirling of Keir, 20th March 1586.—Keir Charters.*

939. COUNTER SEAL OF THE LAST. *Plate VIII. fig. 3.*

A very handsome monogram of the name and title, under a ducal coronet; the background ermine.

940. STUART, ESME, FIFTH DUKE OF LENNOX.

A very pretty signet. Lennox, etc., as in the last. A palm branch at each side of the shield, and a coronet above. Engraved on stone, and mounted with silver in a handle of Egyptian agate.— *In the possession of H. Howard, Esq., Greystoke Castle.*

941. STUART, WILLIAM, DESCENDED FROM STUART OF DARNLEY, COMMENDATOR OF PITTENWEEM. *Plate V. fig. 7.*

Quarterly: first and fourth, three fleurs-de-lis within a bordure engrailed, for D'Aubigny; second and third, a fess chequé within a bordure charged with eight buckles, for Stuart of Darnley. Crest, on a helmet with mantling, a wolf's head placed between the letters w - s. A ribbon issues from the wreath, but no motto appears on it. Supporters, two wolves.

" ✠ WILIEL. STUART COME A. D. PITTENVE. DNI REGAL D. KERYMUR ET BARO. D. ALDBAR." —*Appended to an Assignation to David Lindsay of Edzell, 30th May 1584.— Communicated by W. Fraser, Esq.*

942. STUART, W . . . (?)

A bend surmounted by a fess chequé within a bordure of the same. Crest, on a helmet with mantling. A thistle and rose both slipped and leaved, in saltire. Motto above, "JUVANT ASPERA PROBUM."—*From a triangular steel seal in the British Museum.*

The remaining sides of the seal are occupied respectively with the above crest and the initials w - s. It is a neat example of seventeenth-century work, and doubtless should be assigned to a Stuart, but to which of that numerous family it is perhaps now impossible to say.

943. STUART, RICHARD.

A fine seal of fourteenth-century work. A man sitting in a galley, sailing on the water. A banner at the stem and stern appears to bear two coats quarterly: first and fourth, a lion rampant; second and third, chequé.

" ✠ S' RICARDI SENESCALLI MILITIS." Detached Seal.—*H.M. General Register House.*

944. STUYCKLAND (?), ELIZABETH, Daughter of Andrew Stuyckland.
 Three birds passant.
  " s' elizabeth s . . de."—*Appended to a Charter, a.d. 1368.—Crawford and Balcarras Charters.*

945. STUYESE (?), JOHN.
 A stag's head cabossed, not on a shield; between the antlers a cross.
  " s' johis de stuyese." (?)—*Appended to the same Homage as No. 48, a.d. 1296.—H.M. Record Office.*

946. SUTHERLAND, WILLIAM, EARL of.
 Three mullets. The shield in centre of tracery.
  " s' will. comitis sutherland."—*Appended to an Instrument relating to the Ransom of David II., 5th October 1357.—H.M. Record Office.*

947. SUTHERLAND, ALEXANDER, of Duffus.
 Three cross-crosslets fitchée, and in chief as many mullets.
  " s' alexandri suderlande."—*Appended to the same Instrument as No. 208.*

948. SUTHERLAND, WILLIAM, of Duffus.
 Per fess; the upper part per pale; in the dexter, three mullets, for Sutherland; sinister, as many cross-crosslets fitchée, for Cheyne of Duffus. In the lower part a boar's head, for Chisholm.
  " s' vilelmus suderland."—*Appended to Charter by William Sutherland of Duffus to Hew Lord Fraser of Lovat of the lands of Quarrelwood, 22d October 1540.—Lovat Charters.*

949. SUTHERLAND, MALCOLM.
 A fess between three mullets. Foliage at the top and sides of the shield.
  " s' malcome de sutherlande."—*Appended to Procuratory of Resignation of the lands of Mulquhaich, Drumornie, etc., co. Inverness, a.d. 1476.—Cawdor Charters.*

950. THOMSON, ALEXANDER.

A stag's head cabossed, and on a chief, a crescent between two mullets.

" s' ALEXAN. THOMSONE."—*Appended to Charter by Alexander Thomson, burgess of Edinburgh, and Margaret Lumsden his spouse, of some land and a tenement in Edinburgh, to Joseph Paterson, 10th July 1591.—Edinburgh Charters.*

THOR, LONGUS. *Vide* LONGUS.

951. TOFTIS, INGRAMI DE.

A fleur-de-lis, not on a shield.

" s' INGRAMI DE TOFTIS."—*Appended to the same Homage as No. 48, A.D. 1296.— H.M. Record Office.*

952. TOFTIS, ROBERTI DE.

A wheel or flower ornament, not on a shield.

" s' ROBERTI D. TOFTIS.—*Appended to the same Instrument as the last.*

953. TOK, OR TOKE, JOHN.

A cross; foliage at the top and sides of the shield.

" s' JOANNIS TOK."—*Appended to the same Instrument as No. 702.*

954. TORRIE, GILBERT.

A falcon standing, not on a shield; in the background three stars.

" s' GILBERTI DE TORRI."—*Appended to the same Homage as No. 48, A.D. 1296.— H.M. Record Office.*

955. TRAIL, JAMES.

A chevron between two mascles in chief, and a trefoil slipped in base.

" s' JACOBI TRAYLE."—*Appended to the same Instrument as No. 478.*

956. TRAIL, JOHN, ONE OF THE BAILIES OF ST ANDREWS.

Two trefoils slipped in chief, and a mascle in base.

" s' JOHANNE TRAIL."—*Appended to a Sasine dated 1523.—St. Andrews Charters.*

957. TULLOTH, ARTHUR.

A boar's head contourné erased; in chief two mullets.

" s' ARTHUR DE TULLOTH."—*Appended to the same Instrument as No. 271.*

958. UMFRAVILL, INGRAM.

An orle, ermine, with a label of three points.

"S' INGRAMI DE UMFRAVILLE."—*Appended to Submission of the Competitors for the Crown of Scotland, A.D. 1291.—H.M. Record Office.*

959. VAUS, ROBERT.

A bend; in the sinister chief a lozenge.

"ROBERTI VAUS."—*Appended to an Indenture, dated 13th August 1500.—Edinburgh Charters.*

960. VAUS, JANET.

On a bend three stars; foliage at the top and sides of the shield.

"S' JONETA VAUS."—*Appended to a Charter by Janet Vaus, wife of William Hardyn, burgess of Elgin, to Alexander Paterson, burgess of Inverness, of lands and tenements at Inverness, 7th March 1569.—Local Charters.*

961. VEITCH, ANDREW, PORTIONER IN STEWARTON.

Three bulls' (cows?) heads erased.

"S' ANDRE WICHE."—*Appended to Charter by Margaret Hay, daughter of Andrew Hay, in Ettlesham, with consent of Andrew Wiche, portioner in Stewarton, to John Murray, 28th August 1605.—Blackbarony Charters.*

962. VESCI, LADY ALICIA DE.

Oval-shaped seal. A full-length front figure of a lady, holding in her right hand a shield bearing a cross fleury, and in her left, in front, a cross; above the left shoulder is a shield, vairy. The inscription is lost.

963. COUNTER SEAL OF THE LAST.

A tree, or wheat-sheaf, on which is suspended two shields in the centre, and one on each side. The upper shield bears the cross of the Vescis, and the lower one three garbs for Chester; the dexter shield bears a lion rampant; the shield on the opposite side is lost, as is also the inscription.—*Appended to a Charter by the Lady Alicia de Vesci to the Monks of St. Cuthbert at Durham, probably about the end of the thirteenth century.—Dean and Chapter of Durham.*

964. VESCI, WILLIAM DE.

Three sprigs of vetches.—*Appended to Confirmation by W. de Vesci.—Dean and Chapter of Durham.*

965. VETERIS CASTRI "AVOTE."

An eagle passant to the sinister, not on a shield; the back ground ornamented with stars and a crescent.

" ✠ AVOTE VETERIS CASTRI."—*Appended to a Charter of some land to the burgh of Inverness, by Edona Vetcris Castri " dna ejusdem," with consent of Avote, her daughter and heiress, and that of the Chapter of the Cathedral of Elgin, and the Convent of Inverness, A.D. 1381.—Inverness Charters.*

966. VETRIPONT, WILLIAM.

On a pear-shaped shield, three lions rampant; in honour point a mullet of eight points. At each side of the shield is also one of the same.

" SIGILLUM WILLELMI DE VETERIPONTE."

967. COUNTER SEAL OF THE LAST.

Two demi-lions rampant respecting, not on a shield.

" SIGILLUM SECRETI."—*From General Hutton's Collection. For a detailed account of the family of Vipont, see "Journal of the Archæological Institute," vol. xiii. p. 65; and a notice of this seal will be found in "Catalogue of Antiquities, etc., exhibited in the Museum of the Archæological Institute of Great Britain, at Edinburgh, 1856," p. 91. Edin. 1859.*

966   967.

968. VIPOND, JOHN.

Couché; sinister, six annulets, three, two, and one. Helmet only above the shield, which is in centre of fine pointed tracery.

" SIGILLUM JOHIS WIPOND."—*Appended to a Receipt for five pounds for a war-horse; not dated, but certainly of the 14th century.—Rev. W. Greenwell, Durham.*

X

969. VOLLUM, or WOLLOME, ALEXANDER.

Two bars dancettie.

" s' ALEXANDER WOLLEME."—*Appended to Precept of Sasine in favour of John Lyon and Eufemie Young his spouse, of the fourth part of the lands of Polgorroch, co. Forfar, 24th May 1525.—Crawford and Balcarras Charters.*

970. WAAS, JOHN, of MANNYE.

On a bend, three mullets. Foliage at the top and sides.

" s' JOHANNIS WAAS DE MANE."—*Appended to Charter by John Waas of Many to Sir David Lindsay of Edzell and Jonet Grey his spouse, of the lands of Awchnavis, etc., in the barony of Craigs, co. Forfar, 7th March 1537.— Crawford and Balcarras Charters.*

971. WALDEVE.

A wheel ornament.

" SIGILL. WALDE. FIL. EDWARDI."—*Appended to Confirmation of the Church of Edulingham to St. Cuthbert at Durham.—Dean and Chapter of Durham.*

972. WALDEVE.

Oval-shaped seal. A fleur-de-lis, not on a shield.

" SIGILLUM WALDEVE."—*Appended to a Charter by " Walter filii Uchtred de Reston" of some of his lands at Reston to the Monks of St. Ebbe's at Coldingham.—Dean and Chapter of Durham.*

973. WALLACE, LAMBERT, of SEWALTON (SHEWALTON).

A lion rampant contourné; in the dexter chief a stag's head cabossed.

" s' LAMBART WALLIS."—*Appended to Reversion of an annual-rent from the lands of Sewalton, in the lordship of Fullarton, bailliery of Kirk Stewart, and sheriffdom of Ayr, by Lambert Wallace to John Fullarton of Fullarton, 20th May 1493.—David Laing, Esq.*

974. WARDLAW, SIR HENRY, of WILTON, KNIGHT.

Quarterly: first and fourth, three mascles for Wardlaw; second and third, three water-bougets for Vallange. Crest, resting on the shield, without any wreath, a boar's head and neck couped.

" s' HENRICI DE WARDLAW."—*Appended to Charter by Henry Wardlaw to John, son of Robert, of a tenement in St. Andrews, 10th February 1444.—St. Salvator's College Charters.*

975. WARDLAW, HENRY, of Kilbaberton.
On a fess, between three mascles, a mullet inter two crescents.
" s' henrici vardlav."—*Appended to Sasine of lands of Howburn, co. Peebles, 14th February 1530.—Black barony Charters.*

976. WARDLAW, JOHN.
A chevron between two lozenges in chief and a trefoil slipped in base.
" s' johannis vordlaw."—*Appended to Reversion to Hew Lord Fraser of Lovat; not dated, but probably about A.D. 1560.—Lovat Charters.*

977. WAUCHOP, GILBERT.
A garb; in chief two mullets. Foliage in the background.
" sigillum gilberti wauchop."—*Appended to an Inquisition dated 7th May 1521.— " Panmure Papers."— From General Hutton's Collection.*

978. WEMYSS, SIR JOHN, Knight.
Couché; Quarterly: first and fourth, a lion rampant, for Wemyss; second and third, the same for Inchmartin. Crest, on helmet with mantlings, a swan's head.
" s' johanis vemys d. eodem."—*Appended to Confirmation of Kilmukee Easter, co. Fife, to John Spens, A.D. 1544.— Communicated by W. Fraser, Esq.*

979. WEMYSS, SIR JOHN, Knight, created Earl of Wemyss, A.D. 1633. *Plate V. fig. 4.*
Quarterly, as in the last. Without crest, etc. The initials, s· (Sir John Wemyss) around the shield.
" s' d. johannis wemyss d. eod. militis."—*Appended to Charter of Kilmukee Easter, co. Fife, to David Wemyss, A.D. 1616.— Communicated by W. Fraser, Esq.*

980. WEMYSS, ALEXANDER, of Lathocar.
Per pale dexter a lion rampant, for Wemyss; sinister, a saltire.
" s' alexandri weemys."—*Appended to an Instrument dated 20th October 1511.—St. Leonard's College, St. Andrews.*

981. WEMYSS, JAMES, of Lathocar, son of the above Alexander Wemyss.

Quarterly: first and fourth, a lion rampant, for Wemyss; second and third, a saltire couped.

"s' magis jami weems."—*Appended to a Charter by James Wemyss, A.D. 1557.— St. Leonard's College, St. Andrews.*

982. WEMYSS, THOMAS, of Riras.

Quarterly: first and fourth, a lion rampant, for Wemyss; second and third, a bend for Bisset. Crest, on helmet with mantlings, a swan's neck.

"s' thome de wem."—*Appended to Charter by Thomas Wemys of an annual-rent, 1455.—St. Salvator's College Charters.*

983. WHITELAW, DAVID.

On a chevron, between three boars' heads and necks erased, a mullet.

"s' david quhytlaw."—*Appended to Charter of Deanstone, 20th October 1511.— Blackbarony Charters.*

984. WHITELAW, JOHN.

A chevron between three boars' heads and necks erased.

"s' johannis de quhytlaw."—*Appended to the Perambulation of the lands of Adam Forman, A.D. 1430.—Dean and Chapter of Durham.*

985. WHITELAW, MARIAN.

A chevron between three boars' heads erased; a thistle and rose ornament at the top and sides of the shield.

"s' marian quhytlaw."—*Appended to a Charter by James Roberson, merchant and burgess of Musselburgh, of some lands to the burgh of Musselburgh, 4th June 1633.—Musselburgh Charters.*

986. WILLIAM, Parson of Hunnum, in Teviotdale.

A bird (dove?) passant to the sinister, not on a shield.

"sigillum willi. [parsonne (?)] d. hun."—*Appended to Composition between the Abbey of Melros and William, parson of Hunum, regarding the teinds of Hunum, A.D. 1185.—Melros Charters.*

987. WILLIAMSON, ROBERT, of Muriestion.

Per pale; dexter, a thistle stalked and leaved; sinister, a saltire cantoned in chief and base with a boar's head, and in the flanks with a mullet.

"s' magis. roberti williamsone."—*Appended to Precept of Sasine of Saudersdear, in Haddington, A.D. 1610.—Colston Charters.*

988. WINCHESTER, DAVID, one of the Bailies of St. Andrews.

A griffin, or other fabulous animal, supporting a shield, bearing on a chevron, between two birds (parrots ?) respecting. pecking at fruit on a tree growing from the base, three mullets.

" s' david vynchester."—*Appended to Sasine of an annual-rent of twenty shillings, from a tenement in St. Andrews to the Chaplain and Choristers of St. Andrews. 16th February 1724.—St. Andrews Charters.*

989. WINCHESTER, GEORGE.

Precisely the same design as the last ; the only difference is in the first name.

" s' georgi vynchester."—*From the Collection of General Hutton, who states that the " seal was found in a field near St. Andrews."—" Sigilla," p. 164.*

990. WINCHESTER, THOMAS.

A lion rampant, not on a shield.

" s' thome de winchester."—*Appended to Homage Deed, 1295.—H.M. Record Office.*

991. WISHART, ALEXANDER, Sheriff-Depute of Elgin.

A boar's head couped in chief ; two keys, wards outwards, in base.

" s' alexandri wichard."—*Appended to Sasine of the lands of Moy, by Alexander Ogilvy, brother of the late Patrick Ogilvy of Moy, 15th May 1505.—Cawdor Charters.*

992. WOOD, ANDREW, of Strathtyrum.

A tree erased ; on the dexter, a cross-crosslet issuing from a crescent. Foliage at the top and sides of the shield.

" s' andree wode de strf."—*Appended to Charter by Andrew Wood, of an annual-rent from the lands of Strathtyrum, to Mr. John Rutherford, Provost of the College of St. Salvator, and Canons of the same, 6th August 1577.—St. Salvator's College Charters.*

993. WOOD (or VOLT or VOD), DAVID, " in the Craig."

On a bend, three lozenges ; in the sinister chief a cross-crosslet fitchée.

" s' davit vold."—*Appended to an Agreement regarding the lands of Walterston, in sheriffdom of Forfar, A.D. 1528.—Crawford and Balcarras Charters.*

994. WONMUSE, HENRY.

A hunting-horn stringed, not on a shield.

"s' HENRICH DE WONMUSE."—*Appended to the same Homage as* No. 48, A.D. 1296.
—*H.M. Record Office.*

995. WRIGHT, JOHN.

Two escutcheons in chief and a cross-crosslet issuing from a crescent in base.

"s' JOANNIS WRYCHT."—*Appended to a Charter by John Wright to John Cameron,*
29th January 1437.—St. Leonard's College Charters.

996. YETHAM, WILLIAM OF.

A stag's head cabossed, not on a shield; between the antlers the Saviour on the
cross.

"s' WILLELMI DE YETHAM."—*Appended to the same Homage as* No. 48, A.D. 1296.
—*H.M. Record Office.*

997. YOUNG, JOHN, BAILIE, BURGESS OF EDINBURGH.

Three piles.

"s' JOHANE ZOUNG."—*Appended to Letter of Reversion by John Young to John*
Pennicuick, Younger of that ilk, and Effam Bruce his spouse, with consent of
John Pennicuick, Elder of that ilk, and Marion Hamiltone his spouse, of the
lands of Sleppersfield, etc. etc., in barony of Broughton, co. Peebles, 18th July
1578.—In possession of David Dickson, Esq.

998. YOUNG, SIR PETER, OF SEATON, KNIGHT, TUTOR (UNDER GEORGE BUCHANAN)
OF KING JAMES VI.

A small signet. Three piles, each charged with an annulet. Crest, on helmet with
mantlings, a crescent.—*From a Letter to King James VI., dated 19th June*
1609. Sir James Balfour's Collection, Advocates' Library.

999. YOUNG, WILLIAM.

Three piles, each charged with an annulet; in base an escallop shell.

"s' WILMI ZONG."—*Appended to a Sasine dated 1521.—St. Andrews Charters.*

ECCLESIASTICAL SEALS.

SEALS OF THE BISHOPS OF SCOTLAND.

BISHOPS OF ST. ANDREWS.

1000. EARNALD, Abbot of Kelso, consecrated Bishop of St. Andrews, a.d. 1160.
Plate IX. fig. 6.

    A full-length front figure of a bishop, robed and mitred, his right hand raised
bestowing the benediction, his left hand holding the crosier. The inscription
is rather broken, but from the remaining letters has evidently been " SIGILLVM
EARNALDI DEI GRACIA SCOTTORVM EPI."—*Appended to Confirmation of the
tenths and ecclesiastical rights of the Church of Edenham, the Chapel of
Newton and others, to St. Cuthbert's at Durham, circa a.d. 1180.—Dean
and Chapter of Durham.*

1001. ROGER, Son of Robert Earl of Leicester, from whom the noble house of
Hamilton descends; Lord Chancellor in a.d. 1178; elected Bishop of
St. Andrews circa 1190; died 1202. Plate IX. fig 1.

    A figure of a bishop (?) sitting, holding in the right hand a sceptre or rod, termi-
nating in a fleur-de-lis, the left holding a book in front.

    " ROGERVS DEI GRACIA ELECTVS SANCTI ANDREE."—*Appended to " Charter presenting
Galfrid de Pert to the Church of Rossinclerach by James de Pert, circa
a.d. 1188."—From General Hatton's Collection.*

1002. ROGER, the same person as the last. Plate IX. fig. 3.

    A full-length figure of a bishop in profile, standing on a crescent, in episcopal
vestments, his right hand bestowing the benediction, his left holding the
crosier.

" ROGERUS DEI GRACIA SCOTTORUM EPISCOPUS."—*Appended to an Instrument confirming the rights and liberties of the Church of Coldingham, A.D. 1193.— Dean and Chapter of Durham.*

1003. DAVID BERNHAM, A.D. 1233.

A full-length figure of a bishop in episcopal vestments; profile to sinister; at the dexter side is a crescent and an estoille. The inscription is entirely lost.

1004. COUNTER SEAL OF THE LAST.

An antique gem. Nymphs mocking Silenus.

" MEMENTO DOMINE DAVID."—*Appended to a Letter of the Bishop to the Prior of Coldingham.—Dean and Chapter of Durham.*

1005. ABEL, CONSECRATED A.D. 1254; DIED THE SAME YEAR.

A figure of a bishop similar to the preceding.

" S' ABEL DEI GRA. EPI SCI ANDREA."—*Appended to an Indulgence dated 4th June 1254.—Dean and Chapter of Durham.*

As there were many of these indulgences granted by the Scottish bishops during this and the following century, it may not be uninteresting here to give the form, which is the same in all, except, of course, the date and name of the bishop :—

" Indulgentia xl. diez concessa per Abel dei gra. Epm sci Andree oibz visitantibz feretrum sancti Cuthberti sive Galileau in eccles' Dunhelm' cum orationibz et donis. Dat' apd Dunelm' 4 non' Jun M.CC.LIV."

1006. JAMES BENE (OR BANE), " DOMINUS DE BIRTI," ELECTED A.D. 1328; LORD CHAMBERLAIN 1330; DIED 1332. *Plate X. fig.* 1.

A figure of St. Andrew being bound to his cross by four men, two of whom are on ladders tying his arms; in the niche below is the front head of a bishop. The inscription is much damaged, but seems to have been

" SIGILLUM JACOBI DEI GRACIA EPISCOPI SANCTI ANDREE."—*Appended to a Quitclaim (or receipt) by the Bishop of St. Andrews to the Prior of Coldingham for two hundred marks, 19th February 1329.—Dean and Chapter of Durham.*

1007. HENRY WARDLAW, CONSECRATED AT AVIGNON A.D. 1404; FOUNDED THE UNIVERSITY OF ST. ANDREWS, 1411; DIED 1440. *Plate X. fig.* 3.

An exceedingly fine design. In the centre is St. Andrew extended on his cross, between two shields, each bearing the arms of Scotland. In the richly decorated side niches are figures of angels, and the saints Peter and Paul, in the lower one a bishop kneeling at prayer. At each side is a shield, the dexter one supported on a crosier, bearing on a fess, between three cross-crosslets fitchée, as many mascles or lozenges, the sinister charged with three mascles, for Wardlaw of Torrie, of which family the bishop was second son.

" ṡ HENRICI DEI GRACIA EPISCOPI SCI ANDREE."—*Appended to a Mandate by Henry Bishop of St. Andrews, directed to " Dno Ricardi de Spet decano Xp̄ianitatis de Merse," to institute William Drax Monk of Durham, Prior of Coldingham, presented to the Prior and Convent of Durham 19th June 1419, with certificate that the said decano Xp̄ianitatis inducted the said William Drax in corporal possession " per claves Ecle calium, stolam et corpale, 27 Julii anno sup. diet."—Dean and Chapter of Durham.*

1008. JAMES STUART, SECOND SON OF KING JAMES III., DUKE OF ROSS, ETC., ETC.; CONSECRATED ARCHBISHOP OF ST. ANDREWS A.D. 1497; DIED 1503.

A finely executed round seal, but unfortunately much injured. The arms of Scotland, above the shield a coronet, and the head of the crosier on which the shield is suspended. Supporters, two unicorns. This is among the earliest examples of that animal appearing as the national supporter, but in this instance they are not chained or collared.

There is a ribbon above the shield, on which is the concluding part of the surrounding inscription :—

" ṡ JA. SCI. A[NDREE TOTIUS] SCOCIE PRIMATIS SE. AP. LEGATI DUCIS ROSSIE MARCHIONIS DE ORMODE COM. ARDMANNICH DNI DE BRETHEN (BRECHIN)."—*Appended to a Letter granting permission to Sir John Chapman, chaplain of Ealdham, to take water of the South Esk to supply their mylne pertaining to the chaplaincy, 26th January 1500.—Communicated by W. Fraser, Esq.*

1009. ALEXANDER STUART, NATURAL SON OF JAMES IV., PREFERRED BY THE POPE TO THE ARCHBISHOPRIC A.D. 1509; APPOINTED LORD CHANCELLOR 1511; KILLED AT FLODDEN 1513. *Plate IX. fig. 5.*

As fine a seal as the last, and the same design, except that in this the coronet and ribbon above the shield are omitted. The impression is very perfect.

" ṡ ROTUNDUM ALEXANDRI ACHI EPI. SANCTI ANDREE TOTIUS SCOCIE PRIMATIS SE. AP. LEGATI NAT."—*Appended to Precept of Clare Constat in favour of James Lord Ogilvy, of the lands of Kynnel, etc., 27th May 1506.—Carnegie Charters.*

1010. ALEXANDER STUART. THE SAME PERSON AS THE PRECEDING. *Plate XI. fig. 4.*

A remarkably elegant and well-executed design. On the dexter is a figure of

Y

St. Andrew, his left hand on the cross and a book in his right; on the sinister side the Virgin crowned and holding an open book. Between them is a shield with the arms of Scotland, and above it a cross fleury and the cross of St. Andrew; at each side is a thistle; the whole enclosed in a tressure or scroll ornament.

" S' ALEX. SCI ANDRE ARCHI EPI TOCI REGNI SCOTIE PRIMAT APLICI SEDIS. LEGATI AC COMENDATARII DE DUNFERMLING."—*Appended to Charter of lands in Coddingham to Christina Lumsden, wife of Alexander Ellem, 12th August* 1512.—*Communicated by W. Fraser, Esq.*

1011. ARTHUR ROSS. TRANSLATED FROM THE SEE OF GLASGOW TO ST. ANDREWS A.D. 1684; DEPRIVED 1688; DIED 1704.

Within a niche a figure of a bishop mitred and robed, holding before him, with his left hand, the cross of St. Andrew, in his right hand the crosier; in the lower part of seal, a shield bearing a chevron chequé between three water-bougets; in middle chief, a rose. Above the shield a mitre and mantling.

" SIGILLUM ARTHURI ROSS ARCHI EPISCOPI ST. ANDREE. 1685." And on a ribbon. " SIT CHRISTO SUAVIS ODOR."—*Matrix, Trinity College, Glenalmond, Perthshire.*

1012. SPOTTISWOOD, JOHN, ARCHBISHOP, ST. ANDREWS, A.D. 1615-1639.

A small signet. On a chevron between three trees growing from a mount, a boar's head erased.—*From a Letter addressed to Sir Robert Kerr, dated 31st July* 1623.—*Marquis of Lothian.*

BISHOPS OF EDINBURGH.

1013. ALEXANDER ROSE (A BRANCH OF THE KILRAVOCK FAMILY), CONSECRATED BISHOP OF MORAY 6TH MARCH 1686-7, TRANSLATED TO EDINBURGH SAME YEAR; DEPRIVED 1688; DIED 1720.

A round seal. Per pale; dexter, a saltire; in chief, a bishop's mitre, for the see of Edinburgh; sinister, a boar's head erased between three water-bougets, all within a bordure charged with eight roses, the paternal arms of the Bishop. Above the shield, a mitre with labels. Motto on an inner circle, " PRO DEO ET PATRIA."

" SIGILLUM ALEX. ROSE EPISCOPI EDINBURGENSIS."—*Copper matrix in Trinity College, Glenalmond, Perthshire.*

BISHOPS OF DUNKELD.

1014. RICHARD DE PRÆBENDA, A.D. 1203-1210. *Plate X. fig. 5.*

A full-length figure of a bishop in profile, mitred and robed, his right hand raised giving the benediction, his left holding the crosier.

" SIGILL. RICARDI DEI GRACIA DUNKELDIE."—*Appended to a Composition between the Prior of Durham and St. Andrews, circa 1205.—Dean and Chapter of Durham.*

1015. RICHARD INVERKEITHING, A.D. 1250-1272.

A full-length figure of a bishop, in front, bestowing the benediction.

" SIGILL. R[ICARDI DEI] GRA. EPISCOPI DUNKELDENSIS."—*Appended to an Indulgence similar to that given under No. 1005, A.D. 1254.—Dean and Chapter of Durham.*

1016. WILLIAM.

A figure of a bishop, as in the last, surrounded in the lower half of the seal by a border of six escallop shells.

" [S' WIL]LELMI DI GRA. [DUNKELDEN EPI]."—*Appended to an Indulgence similar to that given under No. 1005, A.D. 1285.—Dean and Chapter of Durham.*

1017. MATTHEW CRAMBETH, A.D. 1289; DIED 1309. *Plate X. fig. 7.*

A full-length front figure of a bishop mitred and robed, his right hand raised bestowing the benediction, his left holding the crosier; in base a lion's head affronté; background diapered.

" S' MATHEI DEI GRA. EPI DUNKELDENSIS."—*From Mr. Doubleday's Collection of Casts, who only gives the reference—" Paris."*

1018. JOHN, A.D. 1356-1365.

Much damaged. The shield on dexter side bears a lion rampant, and on a chief three bells; the sinister shield, two chevrons.—*Appended to Truce between the Kings of England and Scotland (Edward III. and David II.), June 1369.—H.M. Record Office.*

1019. NICOLAS, A.D. 1408-1411. *Plate X. fig. 6.*

A richly designed seal, but unfortunately broken in the lower part, where shields have been. A figure of the Virgin sitting, with the infant Jesus standing in her lap; at each side is a figure of a bishop mitred and robed, in his left hand the crosier; all within Gothic niches, elegantly ornamented; and above is a representation of the Holy Trinity.

" S' NICHOLAI DEI GRA. EPI. DUNKELDEN."—*Appended to a Deed dated A.D 1408.—H.M. Record Office.*

1020. JAMES LIVINGSTON; consecrated a.d. 1476; appointed Lord Chancellor 1483; died the same year.

A fragment of a fine seal of the usual design of figures within niches. The shield in base bears a bend and a wolf's head contourné; but this is not the coat of Livingston, though the seal is appended as the bishop's to an Obligation by four magnates of Scotland, on the part of the King (James III.), for security and indemnity of the Duke of Albany, 2d August 1482.—*H.M. Record Office.*

1021. GEORGE BROWN; consecrated a.d. 1484; died 1514.

A round seal, rather damaged. A full-length figure of a bishop within a Gothic niche; in base, a shield bearing a chevron between three fleurs-de-lis; above the shield a mitre.

" s' georgii brown [epi dunkeld]." Detached Seal.—*H.M. Record Office.*

1022. GAVIN DOUGLAS, Son of the fifth Earl of Angus; consecrated a.d. 1516; died in London 1522.

A round seal, of a similar design to the last. In base, a shield quarterly, first, Angus; second, Abernethy; third, Brechin; fourth, Stuart of Bonkil; on an escutcheon surtout, Douglas.

" s' rotundu gaivini episcopi dunkeld." Detached Seal.—*H.M. Record Office.*

1023. ROBERT COCKBURN, a.d. 1522-1527.

A round seal, in bad condition. A shield bearing three cocks.—*Appended to Truce between England and Scotland, 4th January 1525.—H.M. Record Office.*

1024. GEORGE CRICHTON, a.d. 1527-1543. *Plate XI. fig.* 7.

A round seal, of very good workmanship. Within a niche, having open tabernacle work at the sides, is a figure of St. Columba in episcopal vestments, crowned with the nimbus, holding in his right hand a dove, and in his left the crosier. In base of the seal is a shield bearing a lion rampant, the paternal coat of the Bishop. Above the shield is a mitre.

" s' georgii crechtoun episcopi dunkelde."—*Appended to a Charter by the Bishop and Chapter confirming charter of John Stuart of Arntullie, a.d. 1536, of the lands of Arntullie, etc., in the barony of Dunkeld and sheriffdom of Perth, to the Vicars, Chaplain, and Choristers of the Cathedral of Dunkeld, on payment of a white rose on the Nativity of St. John the Baptist, or three hundred merks when demanded.—Stuart Menzies, Esq. of Culdares.*

1025. JAMES PATON, A.D. 1571-1596.

A round seal. A shield, bearing between six mullets, palewise, a device, perhaps symbolic, of a serpent twined round a cross, or it may be a monogram of the name of the Bishop. The impression is imperfect, rendering the figure very doubtful.

" s' JACOBI EPISCOPI DUNKELDEN."—*Appended to Charter of the lands of Drumbuy, etc., to Thomas Ballantyne, 6th April 1575.—Communicated by G. Logan, Esq. Teind Office.*

1026. PETER ROLLOCK, A.D. 1585-1606, ALSO AN EXTRAORDINARY LORD OF SESSION, 1596-1610.

A full-length front figure of a bishop, holding, with both hands, the crosier. In the base of the seal is a shield bearing a chevron between three boars' heads and necks erased, the paternal coat of the Bishop. The background of seal filled up with foliage.

" . . . EPISCOPI DUNKELDENSIS."—*Appended to a Tack by Mr. Andrew Elder, Vicar of the parish of Menmuir, with consent of the Bishop and Chapter of Dunkeld, to Sir David Lindsay of Edzell, of all and whole the vicarage and small teinds of the lands of Balhall, etc., A.D. 1600.—Crawford and Balcarras Charters.*

1027. ALEXANDER LINDSAY, YOUNGER SON OF LINDSAY OF EVILICK ; CONSECRATED A.D. 1608 ; RENOUNCED EPISCOPACY 1638 ; DIED 1644.

A full-length figure of a bishop in front, his hands closed on his breast, standing on a shield, quarterly, Lindsay and Abernethy ; over all an escallop shell for difference ; all within a bordure. Below the shield a ribbon with motto, which, however, is not legible. The background foliated.

" SIGILLUM ALEXANDRI EPISCOPI DUNKELDENSIS, 1608."—*From a detached seal, dug up in the garden of an old keep called Pithceilis Castle, near Perth.—Communicated by W. de B. M. Galloway, Esq.*

1028. WILLIAM LINDSAY, OF THE FAMILY OF LINDSAY OF DOWHILL, A.D. 1677-1679.

A round seal, having only a shield bearing a fess chequé between a mullet in chief and a base, wavy. Around the shield are the initials $_{w.D.}^{B}$. (William Bishop of Dunkeld).

" GULIELMUS EPS DUNKELDENSIS."—*From the Collection of General Hutton.*

BISHOPS OF ABERDEEN.

1029. JOHN, *circa* A.D. 1204–1207. *Plate* IX. *fig.* 8.

A full-length front figure of a bishop in episcopal vestments, bestowing the benediction, his left hand holding the crosier.

" SIGILLUM JOHANNIS ABERDONENSIS EPISCOPI."

1030. COUNTER SEAL OF THE LAST. *Plate* IX. *fig.* 9.

The Agnus Dei.

" SIGILLUM JOHANNIS EPISCOPI."—*From the late Mr. Doubleday's Collection of Casts, who only gives the reference—" Duchy of Lancaster."*

1031. WILLIAM ELPHINSTONE, BISHOP OF ROSS A.D. 1483; TRANSLATED TO ABERDEEN 1484; LORD CHANCELLOR 1487; LORD PRIVY SEAL 1492; FOUNDED THE UNIVERSITY OF ABERDEEN 1494; DIED 1514. *Plate* X. *fig.* 8.

A round seal, with a well-executed design of three canopied niches; in the centre one is the Virgin with the infant Jesus in her arms, both crowned with the nimbus; in the dexter niche a figure of a bishop, and in the sinister one St. Kentigern, holding in his right hand the fish and ring in its mouth, and in his left the pastoral staff. The figure of St. Kentigern is here introduced to indicate the bishop's connexion with Glasgow, of which diocese he was official, and also Rector of the University. In the lower part of the seal is a shield, bearing the coat of Elphinstone, and above it a mitre; the background diapered.

" S' ROTUND WILLI EPI ABERDONEN."—*Appended to Letter of Collation by William Bishop of Aberdeen, in favour of Alexander Keith, of the Church of Forglen, 28th August 1490.—Communicated by Lord Lindsay.*

1032. WILLIAM ELPHINSTONE, THE SAME PERSON AS THE LAST. *Plate* X. *fig.* 9.

A fine large seal, but broken in the lower part. Oval-shaped. The design is the same as the last.

" S' AUTETICA WILELMI [EPI AB]ERDONEN."—*Appended to an Instrument dated 1501, in the King's College, Aberdeen.*

1033. GAVIN DUNBAR, A.D. 1518–1532.

A full length figure of the Virgin and infant Jesus within a niche, rather rudely executed. In the lower part of the seal is a shield bearing three cushions within the royal tressure; above the shield a mitre.

" S' AUTENTICA GAVINI EPISCOPI ABERDONEN."—*Appended to the Foundation Statutes of King's College, 1530.—King's College, Aberdeen.*

1034. WILLIAM STUART, OF THE FAMILY OF MINTO, A.D. 1532-1545.

A fess chequé of four tracks. A neat round seal. No inscription.—*Appended to Convention between the Ambassadors of England and Scotland, relating to the Castle of Edryngtone, etc., 12th May 1534.—H.M. Record Office.*

1035. WILLIAM GORDON, A.D. 1545-1577.

A rudely executed seal. A triple canopy : in the centre one a full-length figure of the Virgin and Child ; in each of the side ones a figure of a bishop, or saint. in episcopal vestments, with a crosier in the right hand and a book in the left ; in the base of seal a shield. Quarterly : first, three lions' heads erased, for Badenoch ; second, three boars' heads and necks couped, for Gordon ; third, three crescents and royal tressure, for Seton ; fourth, three cinquefoils, for Fraser. Above the shield, a mitre and a scroll, with the words " BENEDICTUS DEUS."

" S' ROTUNDUM VILLELMI GORDON EPISCOPI ABERDONEN."—*Appended to " Tack" of the lands of Kynmock to George Forbes of Kynmock, for eight pounds yearly, A.D. 1576.—Communicated by Lord Lindsay.*

1036. DAVID CUNINGHAM, A.D. 1577-1603.

A large round seal, rather imperfect. Quarterly : first and fourth, a shakefork, in chief a mullet, for Cuningham of Cuninghamhead ; second and third, two garbs, for Mure of Rowellan ; above the shield a clasped book. Supporters. two coneys. There is an inscription at each side of the book, but it cannot be distinctly made out. Below the shield are the words " OUR, OUR," doubtless a contraction of the well-known motto, " OVER FORK OVER."

" S' DAVIDIS CUNINGAMI ABERDONENSIS EPISCOPIS, A.D. 1599."— *Communicated by Cosmo Innes, Esq.*

1037. PATRICK FORBES, OF CORSE, CO. ABERDEEN ; ELECTED A.D. 1618-1635.

A signet. A passion cross between three bears' heads muzzled and couped. Above the shield the initials P · F (Patrick Forbes).—*From a Letter to King James VI., dated 9th September 1619.—Sir James Balfour's Collection, Advocates' Library.*

BISHOPS OF MORAY.

1038. ARCHIBALD, A.D. 1253; DIED 1298.

A full-length front figure of a bishop bestowing the benediction.

"SIGILLUM ARCHEBALDI D[EI GR]A EPI MORAVEN."—*Appended to an Indulgence similar to that given under No. 1005, A.D. 1268.—Dean and Chapter of Durham.*

1039. DAVID STUART, A.D. 1461; DIED 1476 (?).

A shield charged with a fess chequé between two crowns in chief, and a cross-crosslet in base; above the shield a mitre. Inscription, on a ribbon surrounding, "SIGILLUM ROTUNDO. DAVIT EPI MORAVIENS."—*Appended to Confirmation of the lands of Newlands to the Chapel of the Virgin, A.D. 1471.—Inverness Charters.*

1040. WILLIAM TULLOCH.

A fragment of a fine seal. Within a richly decorated niche is a representation of the Trinity, and beneath is a shield bearing on a fess, between three cross-crosslets fitchée, as many mullets, being the paternal coat of the bishop; above the shield is a mitre.

". . . EPI MORAVIEN."—*Appended to Decreet-Arbitral between the Thane of Cawdor and the Baron of Kilravock, whereby a piece of controverted ground lying betwixt the lands of Cawdor and Kilravock is declared to belong to Cawdor, 1st November 1480.— Cawdor Charters.*

1041. ANDREW FORMAN, POSTULATE OF MORAY.

A device, not on a shield, a calf's (?) head; around it the initials A. P. F. (Andrew Forman, Postulate.) Background foliated.—*Appended to Indenture for Marriage between King James IV., and Margaret, daughter of King Henry VII., A.D. 1501.—H.M. Record Office.*

1042. ANDREW FORMAN (SAME PERSON AS THE LAST), BISHOP OF MORAY A.D. 1501; TRANSLATED TO ST. ANDREWS 1514; DIED 1522.

A round seal, rather rude. A representation of the Trinity within a niche; at each side a figure of the Virgin, and a saint. In base a shield quarterly, first and fourth, a chevron between three fishes, haurient, for Forman; second and third, a goose's head, contourné, with a bell fastened to the neck. Supporting the shield are two angels kneeling, and above it a mitre.

"S' ANDRE FORMAN [EPI] MORAVIEN."—*Appended to Precept by the Bishop for infefting Thomas Lord Fraser of Lovat, son and heir of " quondam" Hugh Lord Fraser of Lovat, in the lands of Kiltarlity, 27th April 1502.—Lovat Charters.*

1043. ALEXANDER STUART, SON OF ALEXANDER DUKE OF ALBANY, A.D. 1527-1534.

Much damaged, but has evidently been a fine seal, of a round shape, with the usual design of three saints in niches. The shield in base is quarterly, the first being distinctly Scotland; the other charges cannot be made out. The surrounding inscription appears to be rather a text from Holy Scripture than the name and style of the Bishop.—*Appended to the Gift of Presentation by the Bishop of Moray to Sir Magnus Vaus, of the Chaplaincy of St. Katherine's, 30th November 1536.—Inverness Charters.*

1044. GEORGE DOUGLAS, NATURAL SON OF ARCHIBALD EARL OF ANGUS; CONSECRATED A.D. 1573-74; DIED 1580.

Very imperfect impression. Under a canopy, supported by spiral columns, is a representation of the Trinity, and at the dexter side is a figure of the Virgin, holding the infant Jesus in her arms; at the sinister side is St. Michael vanquishing the Dragon; in base of the seal is a shield, bearing the simple coat of Douglas, and above it a mitre.

" S' GE[ORGII] DOWGLAS IN VIGILAN ENVEIN . . . . " (?)   Detached Seal.—*T. Thomson, Esq.*

1045. ALEXANDER DOUGLAS, PROMOTED TO THE SEE OF MORAY A.D. 1606; DIED 1623.

A signet, rudely executed. A human heart; above the shield the initials A. D.—*From a Letter to King James VI., dated 21st July 1613.—Sir James Balfour's Collection, Advocates' Library.*

1046. JOHN GUTHRIE, PROMOTED A.D. 1623; DEPRIVED 1638.

A round seal of a similar design to No. 1044. In base a shield, quarterly: first and fourth, a lion rampant, for Guthrie; second and third, a garb, for Cumin. On a ribbon above the shield is " JEHOVA PORTIONA," and the date, 1623, is at the sides.

" SIGIL. JOAN GUTHRIE EPISCOPI MORAVIENSIS." —*From an impression on paper in Hutton's " Sigilla." p. 15.*

z

1047. JOHN GUTHRIE. The same person as the last.

A small signet. Quarterly, as in the last. Above the shield the initials B I · M (John Bishop of Moray).—*From a Letter to Viscount Stormonth, dated 27th November 1626.—Marquis of Lothian.*

BISHOPS OF BRECHIN.

1048. ALBINUS, A.D. 1253-1267.

A fine seal. A full-length figure of a bishop in profile, to sinister, standing on a crescent reversed above a fleur-de-lis; at the dexter side is a crescent and star, and at the sinister three points or dots.

"ALBINUS DEI GRACIA BRECHINENSIS EPISCOPUS."

1049. COUNTER SEAL of the last.

The Virgin and Child, and below a monk praying.

"SIGIL. SECRETI A. EPI BRECHINENSIS."—*Appended to an Indulgence similar to that given under No. 1005, A.D. 1256.—Dean and Chapter of Durham.*

1050. WILLIAM, A.D. 1290.

A fine seal, but unfortunately much damaged. The Saviour seated, holding a globe in his left hand, his right raised giving the benediction. The inscription is lost, except the concluding word—"BRECHINENSIS."—*Appended to an Indulgence similar to that given under No. 1005, A.D. 1286.—Dean and Chapter of Durham.*

1051. THOMAS SYDSERF (St. Serf), A.D. 1634; translated to Galloway 1635: deprived 1638; translated to Orkney 1662; died 1663.

A rudely executed design of a representation of the Trinity within a niche formed of heavy masonry in place of the elegant light columns of earlier times : in the lower part of the seal is a shield bearing a fleur-de-lis, and in chief two cinquefoils or mullets.

S' THOME EPISCOPI BRECHINENSIS."—*From an original brass matrix.*

1052. DAVID STRACHAN, CONSECRATED A.D. 1662-1671.

Same rude style as the last. A bishop sitting with uplifted hands. In the lower part of the seal is a shield charged with a stag couchant. The initials ᴮ D·B (David Bishop of Brechin) surround the shield.

"SIGIL. DAVIDIS EPISCOPI BRECHINENSIS."—*Original brass matrix in the Museum of the Society of Antiquaries of Scotland.*

1053. GEORGE HALIBURTON, CONSECRATED A.D. 1678; TRANSLATED TO ABERDEEN 1682; DIED 1715.

Round shape, intended for the shield, bearing on the dexter a figure of a bishop or presbyter, sitting with upraised hands within a niche, no doubt intended for the assumed arms of the see; the sinister side bears, on a bend between three boars' heads and necks erased, as many mascles, the paternal coat of the Bishop.

"SIGILLUM GEORGII EPISCOPI BRECHINENSIS."—*From General Hutton's Collection.*

BISHOPS OF DUNBLANE.

1054. CLEMENT, CONSECRATED A.D. 1233; DIED 1257.

A full-length figure of a bishop, profile to the sinister, in episcopal vestments. On the dexter side is a crescent and on the sinister a star.

"SIGILL. CLEMENTIS DEI GRA. [DUNBLAN]ENSIS."—*Appended to an Indulgence similar to that given under No. 1005, A.D. 1253.—Dean and Chapter of Durham.*

1055. ROBERT DE PRÆBENDA, A.D. 1258-1282.

A fine seal. A figure of a bishop in episcopal vestments, in the usual attitude, within a niche, and at each side, also within niches, a figure of St. Laurence with a gridiron; in the niche beneath, a bishop praying, and at each side a rose.

"S' SECRETI ROBTI DEI GRA. EPI DUNBLANES."—*Appended to an Indulgence similar to that given under No. 1005, A.D. 1260.—Dean and Chapter of Durham.*

1056. MICHAEL OCHILTREE, Dean of Dunblane, a.d. 1430; Bishop 1430-1445.

A round-shaped seal, much injured.   The charges appear to be, on a chevron, between two trefoils, a fleurs-de-lis; the shield is suspended on a crosier, with foliage above.

" s' rotundum michael epi dunblanen."—*Appended to the same Instrument as No. 11.*

1057. JAMES CHISHOLM, eldest Son of Chisholm of Cromlix; consecrated a.d. 1487; resigned 1527; died 1534.   *Plate IX. fig. 7.*

A fine design.   Beneath a canopy, with tabernacle work at the sides, a full-length figure of a bishop in episcopal vestments; in base of the seal a shield, bearing a boar's head and neck erased.

" s' rotunda jacobi epi dublanesis."—*Appended to a Charter of Dispensation to the Earl of Argyle, 13th October 1497.—Breadalbane Charters.*

1058. WILLIAM CHISHOLM, younger Son of Chisholm of Cromlix; consecrated a.d. 1527; died 1564.

Same design as in the last.

s' rotundi. willel. epi dunblanen."—*Appended to a Charter by the Dean and Chapter of Dunblane to Sir James Stirling of Keir, of the lands of Auchinloy, a.d. 1549.—Keir Charters.*

1059. ADAM BELLENDEN, Son of Sir John Bellenden of Auchnoul, a.d. 1615; translated to Aberdeen 1635; deprived 1638.

A round-shaped seal.   A shield bearing a stag's head erased, between two cross-crosslets fitchée.

" sigill. adami episc. dunblanen."—*Appended to a "Tack" (lease) of the teinds of Lenturck, etc., to Robert Gordon of Fechell, 7th February 1620.—Communicated by Lord Lindsay.*

1060. ROBERT LEIGHTON, Dean of the Chapel Royal; Professor of Divinity in the University of Edinburgh, a.d. 1653; consecrated a.d. 1661; translated to Glasgow 1669; resigned 1674; died 1684.

This seal is so sadly broken and defaced, that, but for the interest attached to the person, it might have been excluded.   It proves, however, that Dr. Leighton had a seal as Bishop of Dunblane; it may be much doubted whether he had one as Archbishop of Glasgow.

The design is evidently only the armorial ensigns of the Bishop, but the charges on
the shield are so much flattened that it is impossible to say what they have
been. Above the shield is a mitre and mantlings, and over the mitre a lion's
head erased, as a crest,—an example quite opposed to the "law and practice"
of heraldry, and only to be explained perhaps by the ignorance and confusion of
the period. The few letters remaining distinct, with faint traces of others,
supply the legend thus, " SIGILL. RO. EPISCOPI DUNELANEN. AC DECANNI SACEL.
REG."- *Appended to a Charter by the Bishop, as Dean of the Chapel Royal,
to William Maxwell of Monreith, co. Wigton, of fishings in Kirkcudbright,
with the croft called the fishers' croft, and pasturage, 23d March 1666.—
David Laing, Esq.*

1061. ROBERT LEIGHTON, THE SAME PERSON AS
THE LAST.

A signet. A lion rampant. Crest, on helmet
with mantlings, a lion's head erased.—
*From a Letter to the Duke of Lauder-
dale, no date, but in the year 1673.—
David Laing, Esq.*

Signed

1062. ROBERT DOUGLAS, SON OF ROBERT DOUG-
LAS OF KILMOUTH. DEAN OF THE CHAPEL
ROYAL; CONSECRATED BISHOP OF BRECHIN A.D. 1682; TRANSLATED TO DUN-
BLANE 1684; DEPRIVED 1688; DIED 1716, AGED 92.

A round seal. A shield, quarterly: first and fourth, a human heart, for Douglas;
second and third, a saltire engrailed, the assumed arms of the see of Dunblane; above the shield, a mitre with labels.

" SIG. ROBERTI EPIS. DUMBLANEN AC DECAN SACEL REGII."—*From the original
matrix in General Hutton's Collection.*

1063. ROBERT DOUGLAS. A SECOND SEAL OF THE SAME PERSON AS THE LAST.

A round seal. A shield, bearing on the dexter a saltire engrailed, the assumed
arms of the see of Dunblane; and on the sinister, a human heart ensigned
with an imperial crown; and on a chief, three mullets, for Douglas; above
the shield, a mitre with knotted cord and tassels falling down on each side.

" SIG. ROBERTI EPIS. DUMBLANEN AC DECAN SACEL REGII."—*From the original
matrix in possession of the late P. Chalmers, Esq.*

Both these seals furnish examples of erroneous heraldry, which require to be pointed out. In the first, the heart—no doubt meant for the paternal coat of the Bishop, though it certainly is not so—takes precedence of the arms of the see, contrary to invariable practice. In the second seal this error is corrected by placing them in the dexter side of the shield and impaling the coat of Douglas, which, however, as the Bishop was of the Glenbervie branch of the family, should have been quartered with the embattled cross of Auchinleck.

BISHOPS OF ROSS.

1064. ROBERT, A.D. 1253-1270.

A fragment of a fine seal. A full-length front figure of a bishop in the usual vestments and attitude. The inscription is quite lost.

1065. COUNTER SEAL OF THE LAST.

A full-faced head of St. Winifred (Boniface).

"SCS BONIFATIOUS."—*Appended to an Indulgence similar to that given under No. 1005, A.D. 1255.—Dean and Chapter of Durham.*

1066. ROGER (CALLED ROBERT IN KEITH'S CATALOGUE), A.D. 1284-1304.

A fine design, but unfortunately now much broken. Three niches, in the centre one a figure of a bishop; in the dexter, St. Michael, and in the sinister another saint; in a niche above is the Virgin and Child, and in the base of seal a monk praying, between two shields, each bearing three lions rampant, being the coat of the old Earls of Ross.

"SIGILL. ROGERI DEI GRACIA ROSSENSIS EPISCOPI."—*Appended to an Indulgence similar to that given under No. 1005, A.D. 1304.—Dean and Chapter of Durham.*

1067. JOHN TURNBULL, A.D. 1420-1449. *Plate IX. fig. 4.*

A round-shaped seal. A fine design apparently, though now much damaged. Within a niche is the head of a saint, front face, crowned, with the tiara and nimbus resting on a shield bearing a bull's head caboshed.

"S' JO . . ." is all that remains of the inscription.—*Appended to the same Instrus as No. 11.*

1068. ROBERT COCKBURN, A.D. 1508-1521.

A full-length figure of a bishop, bestowing the benediction, standing within a Gothic niche; the lower part of the seal, usually occupied by the shield, is broken off.

" s' ROBERTI DEI GRATIA EPISCOPI ROSSEN."—*Appended to a Charter by the Bishop to Thomas Lord Fraser of Lovat, of the lands of Kirkton of Kilmorack, 30th August 1515.—Lovat Charters.*

1069. ROBERT CAIRNCROSS, CHAPLAIN TO KING JAMES V.; LORD TREASURER A.D. 1528; BISHOP 1539; DIED 1545.

A round-shaped seal. The design is merely a shield bearing a stag's head couped: above the shield a mitre.

" s' ROBERTI EPI. ROSSII."—*Appended to Charter by the Bishop to Alexander Fraser of Lovat, of the Kirkton lands of Kilverack, A.D. 1545.—Lovat Charters.*

1070. JOHN LESLEY, AUTHOR OF MORE THAN ONE WORK IN DEFENCE OF MARY QUEEN OF SCOTS, AND HER AMBASSADOR TO QUEEN ELIZABETH FROM 1568 TO 1572; COMMENDATOR OF LINDORES, BISHOP, 1566; VICAR-GENERAL OF THE ARCHI-EPISCOPAL CHURCH OF ROUEN, A.D. 1579; BISHOP OF COUTANCES, IN NOR-MANDY, 1593; DIED AT BRUSSELS, 1596.

A round-shaped seal, with a shield bearing a bend, charged with three buckles, tongues bendwise. Surrounding the shield are the initials I · L · E · R (John Lesley, Epi. Rossen), and above it a buckle, the tongue fessewise; below the shield is a ribbon, with a motto which is now illegible.

" s' JOHANNIS E. ROSSEN AC. COMENDATARII DE LUNDORES."—*Appended to Charter by John, Bishop of Ross, to Robert Leslie, and Janet Elphinstone his spouse, of the lands of Ardersier, etc. etc., in Nairn and Inverness, 9th June 1573.—Cawdor Charters.*

1071. DAVID LINDSAY, MINISTER AT LEITH, FATHER-IN-LAW OF ARCHBISHOP SPOTTIS-WOOD, PROMOTED TO THE BISHOPRIC OF ROSS, A.D. 1604; DIED BEFORE 1613.

A small signet. Quarterly: first and fourth, Abernethy; second and third, Lindsay. Above the shield the initials $_{D \cdot L}^{M}$ (Master David Lindsay).—*From a Letter to King James VI., dated 17th June 1607.—Sir James Balfour's Collection, Advocates' Library.*

BISHOPS OF CAITHNESS.

1072. WILLIAM, a.d. 1245 (?)-1261.

A full-length figure of a bishop, in episcopal vestments, bestowing the benediction.

" SIGILLUM WILLI DEI GRA. KATENENSIS."

1073. COUNTER SEAL of the last.

A demi-figure of a bishop, and in base of the seal a monk praying.

" SIGILL. SECRETI W. EPISCOPI CATENE.—*Appended to an Indulgence similar to that given under No.* 1005, a.d. 1255.—*Dean and Chapter of Durham.*

1074. THOMAS MURRAY of Fingask, a.d. 1348; died 1360.

A full-length figure of a bishop in the usual vestments and attitude ; on the dexter side is a shield bearing three mullets, for Murray, and on the sinister another, charged with a lymphad, within a tressure, for the Isles.

" S' THOME DEI GRA EPI CATHENENSIS ET INSULA."—*Appended to Procuration by the Bishops of Scotland for the Ransom of King David II.,* 26*th September* 1357.—*H.M. Record Office.*

BISHOPS OF ORKNEY.

1075. PETER, a.d. 1270, died 1284.

A fine seal. A full-length front figure of a bishop, in episcopal vestments, bestowing the benediction.

" S' PETRI DI GRA ORCADENSIS EPISCOPI."—*Appended to an Indulgence similar to that given under No.* 1005, a.d. 1273.—*Dean and Chapter of Durham.*

1076. JAMES LAW, a.d. 1606; translated to Glasgow 1615; died 1632.

A signet. A bend sinister between a mullet in chief and a cock in base. The initials $_{I\text{-}L}^{M}$ (Master James Law) are above the shield, which is placed between the initials b-o (Bishop of Orkney).—*From a Letter to Sir James Douglas, dated* 24*th March* 1607.—*Sir James Balfour's Collection, Advocates' Library.*

BISHOPS OF GLASGOW.

1077. WILLIAM WISHART, of the family of Pitarrow, Lord High Chancellor, elected to Glasgow, a.d. 1270; consecrated Bishop of St. Andrews 1273; died 1279.

A pretty design of two Gothic arches; in the dexter one a bishop bestowing the benediction; in the sinister St. Kentigern with a fish in his right hand; in a niche below is a bishop kneeling at prayer.

" secretum wischard dei gra episcopi glasguen."—*Appended to an Indulgence similar to that given under No. 1005, a.d. 1276.—Dean and Chapter of Durham.*

1078. WALTER WARDLAW, consecrated a.d. 1368; appointed Cardinal 1381; died 1389. *Plate X. fig.* 4.

A very fine seal, but the impression is now much damaged. Within a richly orna-mented Gothic niche is a full-length figure of the Virgin and Child, and two bishops, one standing and the other kneeling. Two small niches at the sides also contain figures, and in base is a shield, supported by two lions, and bearing on a fess, between three mascles, as many crosses.

" s' walteri [dei gra]cia episcopi glasguensis."—*From the late Mr. Double-day's collection of casts, who gives only the reference " Paris, a.d. 1368."*

1079. JOHN CAMERON, Keeper of the Privy Seal, elect Bishop a.d. 1426; Lord Chancellor 1428; died 1447.

A small octagon seal. Three bars. No inscription. The shield supported on a pastoral staff.—*Appended to a Peace concluded for Governing the Marches, 12th July 1429.—H.M. Record Office.*

1080. JOHN CAMERON. The same person as the last. *Plate IX. fig.* 2.

A round seal; fine design. Beneath a rich Gothic canopy, with tabernacle work at the sides, is a front head of St. Kentigern, mitred and crowned with the nimbus. The bust rests on a shield supported by a crosier, and bears three bars. At each side is the fish with a ring in its mouth.

" s' johannis cameron epi glasguensis."—*Appended to the same Instrument as No. 11.*

2 A

1081. ROBERT BLACADER (FIRST ARCHBISHOP OF GLASGOW), CONSECRATED AT ROME
A.D. 1480; ABERDEEN 1481; TRANSLATED TO GLASGOW 1484; DIED 1508.
On a chevron, three roses. Above the shield, a cross.

"S' ROBERTI GLASGUEN. ARCHI . . . ."—*Appended to Indenture for Marriage between
James IV. and Margaret, daughter of Henry VII., 1501.—H.M. Record
Office.*

1082. JAMES LAW. *Vide No.* 1076.
Oval-shaped seal. A shield bearing a bend sinister between a mullet in the dexter
chief and a cock in the sinister base; scroll ornament at the top and sides
of the shield.

" S' JACOBI ARCHI EPISCOPI GLASGUEN."—*Appended to a Confirmation of the Charter
of Bordlands, 23d September 1616.—Blackbarony Charters.*

1083. ALEXANDER CAIRNCROSS, OF COWMISLIE; CONSECRATED A.D. 1684; DEPRIVED
1687; BISHOP OF RAPHOE, IN IRELAND, 1693; DIED 1701.
A large seal, rather clumsily designed. A fanciful-shaped shield, bearing on the
dexter the fish with a ring in its mouth, the tree, bird, and bell, the well-
known emblems of the city of Glasgow, and also the assumed arms of the see,
impaling a stag's head erased; between the antlers a cross-crosslet fitchée;
in chief a mullet. Above the shield a front demi-figure of a bishop mitred,
between a crosier on the dexter and a cross on the sinister, placed saltirewise.

" SIGILLUM ALEXANDRI CAIRNCROCE ARCHIEPISCOPI GLASGUENSIS;" and in an inner
circle is inscribed, " PRO DEO. REGE ET ECCLESIA SACRA."—*From the original
brass matrix in Dr. Rawlinson's Collection.—Bodleian Library, Oxford.*

1084. JOHN PATERSON, FORMERLY DEAN OF EDINBURGH, PREFERRED TO THE BISHOPRIC
OF GALLOWAY A.D. 1674; TRANSLATED TO EDINBURGH 1679; GLASGOW 1687;
DEPRIVED 1688; DIED 1708.
A round seal. Per pale; dexter, a tree, on which is a bird, and from a branch a
bell is suspended; on a chief, a demi-figure of a bishop; in base, a fish with
a ring in its mouth. These are doubtless intended for the armorial ensigns of
the see, though differing from those usually given to it, which, like some others
of the sees of Scotland, are assumed without authority. The sinister side of
the shield bears the paternal coat of Paterson, viz., three pelicans in their
piety, and on a chief three mullets. Above the shield a mitre, and the motto,
" PRO REGE ET GREGE;" and below it, on a ribbon, " CONSTANT AND TRUE."

" SIG. JOHAN PATERSON ARCHIEPISCOPI GLASGUENSIS, 1687."—*Original copper matrix
at Trinity College, Glenalmond, Perthshire.*

1085. JOHN PATERSON. The same person as the last.

A small signet, very superior in the execution to the larger seal. The marshalling of the shield is the same, but has the crosier and episcopal staff placed behind it, saltirewise, and the motto, " PRO REGE ET GREGE" is on a ribbon below, omitting the second motto.—*Steel seal, in the same Collection as the last.*

BISHOPS OF GALLOWAY.

1086. GILBERT, consecrated a.d. 1235; died 1253.

A full-length figure of a bishop (St. Ninian?), in episcopal vestments, bestowing the benediction.

" S GILBERTI GRACIA DEI CANDIDE CASA EPISCOPI."

1087. COUNTER SEAL of the last.

The Agnus Dei.

" JASPIDISS VIRTUS FUSUM SEDARERNOREM."—*Appended to an Indulgence similar to that given under No.* 1005, a.d. 1248.—*Dean and Chapter of Durham.*

1088. HENRY, elected a.d. 1253; consecrated 1255-1292.

A similar design to the preceding.

" SIGILL. HENRICI CANDIDE CASA EPISCOPI."

1089. COUNTER SEAL of the last.

A figure of a bishop kneeling before another, intended probably to represent the rite of consecration.

" SCE. P. CORDANE TIBI DISSPLICEAR NINIANE."—*Appended to an Indulgence similar to that given under No.* 1005, a.d. 1259.—*Dean and Chapter of Durham.*

1090. THOMAS. Date of consecration and death not ascertained; was certainly bishop in a.d. 1296 and 1304. *Plate XI. fig.* 6.

A full-length front figure of a bishop in the usual attitude. The inscription commences in a singular way, with the name " THOMAS" in a horizontal line across the centre of the seal, and is then continued in the usual way around it.

" DEI GRA. EPISCOPUS CANDIDE CASA."—*Appended to an Indulgence similar to that given under No.* 1005, a.d. 1302.—*Dean and Chapter of Durham.*

1091. ALEXANDER VAUS, of the old family of De Vallibus, a.d. 1426; resigned 1450 (?).

Imperfect impression, apparently a device of an angel with wings expanded, supporting in front a bust or shield. No inscription, but appended as the seal of the bishop to a "Concordia," by the Commissioners for Governing the Marches, 12th July 1429.—*H.M. Record Office.*

1092. HENRY WEMYSS, natural Son of King James IV., a.d. 1526-1541.

A triple niche, containing figures of the Virgin, a bishop, and St. Michael. In the lower part is a shield, but the charges are quite indistinct, as is also the inscription.—*From the Collection of General Hutton, who gives merely the note, "Charter dated 20 Feby. 1538-9."*—*"Cassillis Papers."*

1093. ALEXANDER GORDON, second Son of John, Master of Huntly. Archbishop of Athens; Bishop of the Isles a.d. 1553; translated to Galloway 1558; suspended 1572; Lord of Session 1565; died 1576.

The design similar to the last. The shield in the lower part of the seal is quarterly: first, three boars' heads couped, for Gordon; second, three lions' heads erased, for Badenoch; third, three crescents within the Royal tressure, for Seton; fourth, three frasiers, for Fraser.

"s' ALEXANDRI EPI. CANDIDE CASA."—*Appended to a Charter by the Bishop of Galloway, with consent of the Abbot and Monastery of Tungeland, to Sir John Gordon of Lochinvar, of an annual-rent from sundry lands in the parish of Tungeland and stewartry of Kirkcudbright, 15th April 1564.* — *David Laing, Esq.*

1094. WILLIAM COWPER, a.d. 1614; died 1619.

A mere fragment, unfortunately, of what seems to have been a fine seal, and rather singular design. On the dexter side are clouds, from which issues a hand holding a lighted candle; surrounding it is a ribbon, containing an inscription, which is now quite illegible. A figure of an angel may be traced among the clouds. The lower part of the seal has evidently been occupied by the paternal shield of the Bishop, bearing the same as in the following, between the initials w · c.—*Appended to Charter by the Bishop and Chapter of Galloway to Sir Robert Gordon of Rothemay, of some lands in the stewartry of Kirkcudbright, a.d. 1614.*—*David Laing, Esq.*

1095. WILLIAM COWPER. The same person as the last.

A signet. A chevron, ermine, between three laurel leaves. The initials $\begin{smallmatrix} M \\ W \cdot C \end{smallmatrix}$ (Master William Cowper) around the shield.—*From a Letter to John Murray of Lochmaben, Groom of H.M. Bed-Chamber, dated 10th August 1618.— Marquis of Lothian.*

1096. JOHN PATERSON. *Vide No. 1084.*

A shield, per pale; dexter, a full-length figure of a bishop, holding in his right hand a crosier, the assumed arms of the see, impaling the coat of Paterson, viz., three pelicans in their piety, and on a chief three mullets. Above the shield a mitre with labels. Motto on a ribbon above, "PRO REGE ET GREGE."

" SIG. D. JOAN PATERSON EPIS. CAND. CASE."—*From the copper matrix at Trinity College, Glenalmond, Perthshire.*

BISHOPS OF ARGYLE

1097. ALAN, elect a.d. 1250; died 1262.

A fine seal, with the usual design of a bishop bestowing the benediction. In a circular panel on each side is a front-faced head, and in the niche below is another front head.

" SIGILL. ALANI DEI GRA EPI ERGADIA."—*Appended to an Indulgence similar to that given under No. 1005, a.d. 1255.—Dean and Chapter of Durham.*

1098. ANDREW, a.d. 1304-1329 (?).

The same design as the last.

" S' FRATRIS ANDREE DEI GRA EPI ERGADIE."—*Appended to an Indulgence similar to that given under No. 1005, a.d. 1310.—Dean and Chapter of Durham.*

1099. GEORGE LAUDER, of Balcomy, in Fife. a.d. 1427-1472 (?).

A shield suspended on a crosier, bearing a griffin segreant within the Royal tressure.

" S' ROTUNDUM GEORGII DEI GRA EPI ERGADIE."—*Appended to the same Instrument as No. 11.*

The Bishop of Argyle was not one of the contracting parties to this agreement, but his seal occupies the place of the " sele of a venerabil fader in Christ,

David, Abbot of Cambuskeneth, procurit be William Cranston, burgess and Commissary of Edinburgh," who was one of the contracting parties on the part of the burghs of Scotland.

1100. GEORGE LAUDER, THE SAME PERSON AS THE LAST.

A richly-designed seal; oval-shaped. Within a canopied niche, a full-length front figure of a bishop, in episcopal vestments, crowned with a nimbus, his hands clasped on his breast. In the lower part of the seal, within an arched niche, is a demi-figure of a bishop in front, and at each side is a shield bearing a griffin segreant contourné, within the Royal tressure. Above the canopy is a third shield, placed under a coronet, with the same charge.

" S' GEORGII DEI GRACIA EPI LEGADIEN. LIS."
—Appended to a Dispensation to the Earl of Argyle, 20th November 1455.
—Breadalbane Charters.

1101. JAMES HAMILTON, NATURAL SON OF JAMES LORD HAMILTON, POSTULATE, OR ELECTED TO THE SEE OF GLASGOW, A.D. 1547; ARGYLE, 1558-1575 (?). Plate XI. fig. 2.

A figure of a bishop, bestowing the benediction, standing within a niche having open screen-work at the sides. In the lower part of the seal is a shield bearing, quarterly, Hamilton and Arran; above the shield a mitre.

" S' ROTUNDUM JACOBI EPISCOPI LISMOREN."—Appended to a Charter by the Bishop and Chapter to James Duke of Chatelherault (the Bishop's brother) of sundry lands in Kintyre, 16th May 1556.— Argyle Charters.

1102. ANDREW BOYD, NATURAL SON OF THOMAS LORD BOYD, A.D. 1613; DIED 1636.

A round seal with only a shield bearing a fess chequé; a kind of heavy scroll frame surrounds the shield.

" SIG. AND. EPI LIS."— Appended to a Charter by the Bishop, Dean, and Chapter of Argyle to William Stirling of Auchyll, of some lands in Argyle, 6th November 1629.—Argyle Charters.

BISHOPS OF SODOR, OR MAN, AND THE ISLES.

1103. RICHARD, A.D. 1257-1274. (?)

A fine seal, executed in a bolder style of relief than usual. A full-length front figure of a bishop in the usual vestments and attitude.

["· s' ricar]di epi sodoren manerem et insula."

1104. COUNTER SEAL of the last.

An antique gem, with the device of a chimera.

"· ascende calve as . . . . salve."—*Appended to an Indulgence similar to that given under No 1005, A.D. 1257.—Dean and Chapter of Durham.*

1105. JOHN CAMPBELL, Commendator of Ardchattan, and Elect or "Postulate" Bishop of the Isles, A.D. 1558.

Within a niche a full-length figure of St. Columba, vested in episcopal robes, mitred and crowned with the nimbus, his right hand raised, a dove resting on his left. In the lower part of the seal a shield, between the initials i·c. Quarterly: first and fourth, gyronny of eight, for Campbell; second, a galley for lordship of Lorn; third, a stag's head caboshed, and on a chief a buckle, the tongue fesswise between two mullets, for Calder.

"· s' johis postulate soder ac come. ar[dchattan]."—*Appended to Charter by John, Commendator of Ardchattan to Alexander Campbell of Finis Moir, his brother-german, of five plough-lands of Geddes, in the barony of Mickle Geddes and sheriffdom of Nairn, 10th March 1561.—Cawdor Charters.*

SEALS

OF

ABBOTS, MONASTERIES, ETC., OF SCOTLAND.

1106. ABERDEEN, ADAM TYNYNGHAM, Dean and afterwards Bishop of, A.D. 1382.

Within a double niche, richly ornamented, a full-length figure of the Virgin, before whom is a pot with lilies growing, and a monk holding or presenting her with a scroll, on which is inscribed a word now illegible. In base a shield, bearing a tree between two deers (?) rampant adorse.

"S' ADE DE TYNYGHAM DECANI ABERDON."—*Dean and Chapter of Durham.*

1107. ABERDEEN, THOMAS CARNOTTO, Dean of, and Chancellor of Scotland, A.D. 1342.

A pretty seal, but part is lost. A full-length front figure of a monk, holding, with both hands, a book or a purse. A shield is suspended from a tree, at the sinister side, but the charge is quite defaced.

"S THOME DE CARNOTO . . . DOXIE."—*Dean and Chapter of Durham.*

1108. ABERDEEN, Chapter of the Virgin Mary of Mount Carmel.

The Carmelites were of the order of the Mendicants, deriving their name from Mount Carmel. They were called "White Friars" in Scotland, where they had several establishments; that at Aberdeen is said to have been founded by Philip de Arbuthnot, of that ilk, A.D. 1350.

A double niche; in the dexter a full-length figure of the Virgin holding the infant Jesus in her arms. In the sinister a bishop robed and mitred. In the centre, above, is the Saviour on the cross, and in the base of seal a monk praying.

"SIGILLVM CAPITVLI . . . ."—*Appended to a "Charter granted by the Convent to R. Barron, dated 7 Jany. 1437."*—"*Communicated by Professor Stuart, Mar. Col. 1756."—Hutton's "Sigilla," p. 4.*

1109. ABERDEEN, Prior of the Carmelites of.

A singular design, representing the Resurrection. Beneath a Gothic canopy, a figure of the Saviour with a cruciform nimbus, a cross in his left hand, his right pointing upwards, stepping out of a tomb. All that remains of the inscription is " SIGILL. PRO . . . COCTE."—*Appended to a Charter by the Prior and Convent of the Carmelites of Aberdeen of the lands of Glenfalch to Duncan, son of John, for payment of a silver penny yearly, A.D. 1411.— Marischal College, Aberdeen.—Hutton's "Sigilla," p. 2.*

1110. ABERDEEN, Common Seal of the Carmelites of.

A round seal. A geometrical figure of intersecting triangles, or rather a mullet of five points merely in outline, and the letters M · A · R · I · A in the external spaces. " S' COE. DOME DE ABERDEEN ORDIS CARMELITAE."—*This seal is appended to various charters, etc., in Marischal College, Aberdeen.—Hutton's "Sigilla," p. 1.*

1111. ABERDEEN, Dominicans or Friars-Preachers of.

The Dominicans (first order of Mendicants), called also Friars-Preachers, and Black Friars, were instituted by St. Dominic about A.D. 1215. They had several monasteries in Scotland. That of Aberdeen was founded by King Alexander II.

A full-length figure of St. John the Baptist, holding in his left hand a circular disc, on which is the Agnus Dei, to which his right hand is pointing. In the background are two trees and foliage. The inscription appears to be " SIGILLUM COMMUNE FRATRUM ORDINIS PREDIC DE ABERDEN."—*Appended to a Charter by William de Daulton, brother of the Order, granting to the minister and Trinity Friars of Aberdeen an annual-rent of 13s. 4d. out of his lands at Aberdeen, 30th September 1381.—Hutton's "Seals," p. 117.*

1112. ABERDOUR, Fife. A Nunnery of the order of the Franciscans or Claresses, so called from St. Clare.

The Franciscans (second order of the Mendicants) were instituted by St. Francis, A.D. 1206; known also as Friars-minors, and Grey Friars. There were only two nunneries of this order in Scotland—Aberdour and Dundee,—and scarcely anything is known of their history. They were probably not very numerous in either establishment.

A very imperfect impression. A figure of St. Clare, or the Virgin Mother, holding a lily, with her right hand on her breast, within a niche. " S' COMMUNE . . . ."—*Appended to a Tack of the Sister-lands of Aberdour to James Earl of Morton, A.D. 1560.—Morton Charters.*

2 B

1113. ARDCHATTAN, Argyle. A Priory of the Monks of Valliscaulium, a reformed
Order of the Cistercians, founded in A.D. 1193.

There were only three establishments of this reformed order in Scotland,—Ardchattan, Beauly, and Pluscardine. The first was founded by Duncan Mackoul
A.D. 1230. *Plate XII. fig. 2.*

A very fine seal. A full-length figure of St. John the Baptist, holding the Agnus
Dei encircled with the nimbus. At his right side are two monks kneeling
and praying.

" SIGILL. CONVENTUS DE ARDKATAN IN ARGADIE."—*Appended to a Charter by John,
Perpetual Commendator of Ardchattan, to Alexander Campbell of Flenismoir,
of the four mark lands of Ardache, etc., 10th June 1564.—Argyle Charters.*

1114. ARGYLE, OR LISMORE, CHAPTER OF.

A full-length figure of a bishop giving the benediction ; beneath a canopy with
tabernacle-work at the sides. Very rudely executed.

" S' COMMUNI CAPITULI ECCLESIE LISMOREN," A.D. 1574.—*Breadalbane Charters.*

1115. AYR, PRIOR OF THE FRIARS-PREACHERS OF. (*See* No. 1111.) This house was
founded either by William, Bishop of St. Andrews, or by King Alexander II.,
about A.D. 1230.

An imperfect impression, but it appears to have been a fine design of St. Catherine,
within a Gothic niche, leaning her right hand on a sword, and in her left holding
a wheel, the instrument of her martyrdom ; in lower part a monk at prayer.

" S' PRIORIS FRATRUM ORDINIS PREDICATORUM DE AR."—*Appended to an Instrument
dated A.D. 1406.—Hutton's "Seals." p. 13.*

1116. BANFF, CARMELITES OF. (*Vide* No. 1108.) It is not known by whom this house
for the Carmelites was founded. It was dedicated to the Virgin Mary, and
survived the period of the Reformation till the year 1617, when all its pos-
sessions were annexed to the old College of Aberdeen.

The Salutation of the Virgin, and below is a monk praying. The inscription is very
indistinct, but seems to be " AVE MARIA PLENA GRACIA DOMINUS."—*Appended
to a " Precept of Sasine in favour of Sir William Ogilvy of Dunlugas."—
Hutton's "Sigilla," p. 99.*

1117. BEAULY, ROSS-SHIRE. A Priory of the same order as No. 1113.

The Virgin and Child sitting within a niche; below is a monk praying. Inscrip-
tion not legible.—*Appended to a Tack of the Teinds of Conveth to Hugh Lord
Fraser of Lovat, by Walter Reid, Abbot of Kinloss and Prior of Beauly,
A.D. 1571.—Lovat Charters.*

1118. BRECHIN, WILLIAM CARNEGIE, Preceptor of Maison Dieu. The " Maison Dieu" was an hospital founded by William de Brechin in the thirteenth century for the relief of the poor.

The Virgin and infant Jesus standing in a crescent, and surrounded by a flamboyant aureole.

" S' DNE WILLELMI CARNEGY PRECEPTORIE DE MESONDEU," A.D. 1549.—*Southesk Charters.*

1119. CAITHNESS, Chapter of St. Mary's.

A rude copy of the fine old seal of the Chapter. See No. 988, "*Descriptive Catalogue of Ancient Scottish Seals.*"—*Silver matrix at Trinity College, Glenalmond, Perthshire.*

1120. COLDINGHAM, Berwickshire, SIMON, Prior of. Plate XII. fig. 6. A Priory of the Order of the Benedictines, or Black Monks, which was founded by St. Benedict (Bennet) as early as the fifth century, and was the first and most influential of the monastic orders in Europe. The Priory of Coldingham was founded by King Edgar about 1098, in honour of St. Cuthbert. A nunnery had previously existed at Coldingham, but it is not known that its occupants followed the rule of St. Benedict.

A figure of a monk sitting before a lectern reading.

" SIGILL. SIMONIS PRIORIS DE COLDINGEHAM."—*Appended to Manumission by Simon, Prior of Coldingham, of Hobbne, son of Dencan, and his heirs, circa 1100 — Dean and Chapter of Durham.*

1121. COLDINGHAM, THOMAS DE MELSONBY, Prior of. Plate XII. fig. 4.

The Virgin seated, holding a globe ensigned with a cross patée in her left hand, her right lying on her breast. The name " MARIA" on a scroll at the right side.

" SIGILLUM THOME PRIORIS DE COLDINGHA."—*Appended to Indenture between the Prior and Convent, and Patrick Earl of Dunbar, regarding the Marches of Billie; without date, but certainly before A.D. 1250.—Dean and Chapter of Durham.*

1122. COLDINGHAM, JOHN AYCLIFF (or AKLIFE), Prior of.

A round seal. A figure of a bishop bestowing the benediction ; at his right side a monk standing ; fine tracery surrounds the design.

" SIGILL. MAGRI JOHIS AKLIFE."—*Appended to a Bond by Thomas Purves of Edderham to the Prior of Coldingham and Alexander Home, as security for a debt to them by the said Thomas Purves, 3d June 1407.—Dean and Chapter of Durham.*

1123. COLDINGHAM, WILLIAM DRAX, PRIOR OF.

A pretty design of a double niche and canopies, in the dexter the Virgin and Child, and in the sinister a bishop, both sitting; below is a demi-figure of a monk praying.

"VOS SCI DEI SITIS AMICI MEI."—*Appended to a Lease by the Prior of Coldingham to John and William Manderston, of all his lands of Edderham for five years, 20th May* 1422.—*Dean and Chapter of Durham.*

1124. COLDINGHAM, JOHN OLL, PRIOR OF.

A similar design to the last, and equally fine work. Above the canopy in this instance is an eagle (the eagle of St. John (?) in allusion to the name).

"S' JOHIS OLL PRIORIS DE COLDYNGHAM."—*Appended to a Patent to Sir Alexander Home, Knight, appointing him Bailie and Governor of the lordship and lands of Coldingham, 13th May* 1442.—*Dean and Chapter of Durham.*

1125. COLDINGHAM, COMMON SEAL OF THE MONASTERY OF. *Plate* XII. *fig.* 5.

A fine seal, in excellent preservation. A full-length figure of the Virgin, crowned with a royal crown and nimbus, with the infant Jesus, also crowned with the nimbus, in her arms; the background ornamented with foliage.

"SIGILLUM COMUE MONASTERII DE COLDINGHAM." Detached Seal.—*Communicated by G. Logan, Esq., Teind Office.*

1126. COLDSTREAM, BERWICKSHIRE. A Nunnery of the Cistercians, founded by Cospatrick Earl of March and his lady, about A.D. 1160. The Cistercians, or Bernardines, an order following the rule of St. Benedict, was instituted in France about the end of the eleventh century; they had several flourishing houses in Scotland. *Plate* XV. *fig.* 5.

A fish (salmon ?) biting at a line, between a star-fish (or estoile), a crescent, and a quatrefoil on the dexter, and a wheel and quatrefoil on the sinister side.

"SIGILL. SCE MARIE DE CALDESTREM."—*Appended to an Indenture between the Prior (John) and Convent of Durham, and Mariori' prioris and Convent of Coldstream, dated 10th October* 1419.—*Dean and Chapter of Durham.*

1127. CROSSRAGUEL, AYRSHIRE. *Plate* XII. *fig.* 8. A Monastery of the Cluniac Order, following the rule of St. Benedict (*see* No. 1120), founded by Duncan, son of Gilbert Earl of Carrick, A.D. 1244.

The Virgin with the infant Jesus in her arms, sitting within a richly designed niche, the background diapered; in base is a group of four monks praying.

"SIGILLU. BEATE MARIE DE CROSREGMEL."—*Appended to a Charter by the Abbot and Convent, dated* 1546.—*Cassillis Papers.—General Hutton's Collection.*

1128. DEER, ABERDEENSHIRE. *Plate* XI. *fig.* 5. A Monastery of the Cistercian Order (*see* No. 1126), founded by William Cumin, Earl of Buchan, A.D. 1218.

A full-length figure of an abbot, with a crosier in his right hand, and a book in his left, standing on a crescent reversed; the background richly diapered with foliage.

" SIGILLUM COMMUNE MONASTERII DE DER."—" *Appended to Procuratoric of Resignation of the lands of Deir, in favour of Robert, Commendator of Deir, and George Earl Marischal,* A.D. 1587."—*Hutton's* " *Sigilla,*" p. 80.

1129. DRYBURGH, TEVIOTDALE, JAMES OGILVY, COMMENDATOR OF. A Monastery of the Praemonstratenses, an Order of the Canons regular of St. Augustine, founded in the twelfth century by St. Norbert, a German archbishop of Magdeburg. There were six houses of this order in Scotland. Dryburgh was founded by Hugh Morville, Constable of Scotland, and his wife Beatrix, *circa* 1150.

Within a Gothic niche a figure of the Virgin and Child; in base a shield quarterly: first and fourth, a lion passant gardant, for Ogilvy; second and third, three crescents, one and two, for Seton (?).

" S. M. JACOBI OGILVI (?) COMMENDATARI MONASTERI DE DRIBURGH."—*H.M. Record Office.*

1130. DUMFRIES. A Convent of the Franciscans (*see* No. 1112), founded by Lady Devorgilla Balliol, *circa* 1280.

A front figure of St. Francis, his right hand raised, and a small figure of a monk kneeling beside him.

" S. GARDANE FR[ATRUM] DE DRUMFRE."—*Appended to the* " *Confirmation of an Indenture between Beatrix Countess of Errol and the Minorites of Dundee,* 11th *June* 1490."—*From the original at Slains Castle.*—*Hutton's* " *Sigilla,*" pp. 56, 58.

1131. DUNDEE. A CONVENT OF THE FRANCISCANS OF (*see* No. 1112). Founded by Lady Devorgilla Balliol.

A similar design to that of Dumfries, but in this instance the hands of St. Francis are both raised.

" SIGILLUM GARDIAN [FRATRUM MINOR]UM DE DUND."—*Appended to the same Instrument as the last.*—*Hutton's* " *Sigilla,*" p. 57.

1132. DUNDRENNAN, GALLOWAY. A Cistercian Monastery (*see* No. 1126), founded by Fergus, Lord of Galloway, A.D. 1142.

A hand vested, issuing from the dexter, holding a crosier. In the background is an estoile of five points, a star of six, and a crescent.

"CONTRA S' DE DUNDRAYNAN."—*Dean and Chapter of Durham.*

1133. DUNFERMLINE, FIFE. A Monastery of the Order of the Benedictines (*see* No. 1120), founded by Malcolm Canmore and his Queen, Margaret, *circa* 1060, in honour of the Holy Trinity.

A font, within the porch of a church, and above it, issuing from the circumference of the seal, is a dexter hand, the first three fingers extended (symbolical of the blessing of God).

"SIGILLUM SANCTE TRINITATIS."—*Appended to a Deed by the Abbot (R) and Convent, about* A.D. 1200.—*Dean and Chapter of Durham.*

1134. DUNFERMLINE, JAMES BETHUNE (OR BETON), YOUNGEST SON OF BETHUNE OF BALFOUR, ABBOT OF, AFTERWARDS BISHOP OF GLASGOW, ARCHBISHOP OF ST. ANDREWS, ETC. ETC.

A small signet. On a shield, between the letters I · B, an otter's head erased.— *Impressed on paper, being a Receipt, dated 31st January* 1506, *in the Archives of the burgh of Peebles.*—*Hutton's "Seals,"* p. 5.

The otter's head here is a portion only of the *maternal* coat of the Abbot, and may perhaps have been assumed to denote his cadency. Subsequently he carried his full paternal coat, Bethune and Balfour, quarterly. *See* Nos. 879-880 "*Descriptive Catalogue of Ancient Scottish Seals.*"

1135. DUNGLAS, EAST-LOTHIAN, THE COLLEGIATE CHURCH OF. Founded by Sir Alexander Home, Knight, in 1450.

A rudely executed seal. The Virgin, with the infant Jesus in her arms, standing within a kind of Gothic niche.

"S' COMMUNE ECCLESIE COLLEGIATE DE DUNGLAS."—*Appended to an Instrument dated* A.D. 1604.—*G. Logan, Esq., Teind Office.*

1136. EDINBURGH. St. Catherine of Sienna. A Nunnery of the Dominican Order (*see* No. 1111), founded by Lady Janet Hepburn, daughter of Patrick, first Earl of Bothwell, and widow of George, fifth Lord Seton, who fell at Flodden in 1513. *Plate* XII. *fig.* 3.

A figure of St. Catherine crowned with the nimbus, holding in her right hand a crucifix, and in her left a heart; beneath her feet is a figure of the devil. The background foliated.

"S' CONET . . . E EDIBURCH SORORU. SCE. KATHERIE DE SEINS."— "*Charter be Christiane Ballandine, prioriss of ye Systers of Seyus, to Fr. Lewis Ballandine* 15 *Feb.* 1562."—*Hutton's "Seals,"* p. 130.

1137. EDINBURGH. St. Mary's in the Fields. A collegiate church outside of the walls of the city, the site of which is now occupied by the University.

The Virgin and infant Jesus standing within a niche. Foliage at the sides.

"s'[coe. co]llegii beate [mari]e de cam."—*From the Collection of General Hutton.*

1138. EDINBURGH. Trinity College, or Hospital, founded by Mary of Gueldres, Queen of James II., A.D. 1462, in honour of the Holy Trinity, the Virgin, etc. *Plate XI. fig. 3.*

Within a Gothic niche the usual representation of the Trinity. In base part of the seal is a shield bearing a chevron between three buckles, for Bonkil or Bonele, being the arms of Sir Edward Bonele, who was the first Provost of the College.

"s' capituli collegiate ecclie. sce. trinitatis ppe edynburgh."—*Appended to an Instrument uniting the Vicarage of the parish Kirk of Wemyss to the Provostry of Trinity College, 19th September 1502.—City of Edinburgh Charters.*

1139. EDINBURGH. Trinity College. The same Institution as the last.

A perfect and very interesting seal, representing the College Church. Above are the words "scta. trinitas unus deus," amid rays of glory. In the lower part of the seal is a shield bearing Scotland, impaling Gueldres, the arms of the foundress; above the shield an arched crown. This represents the seal as ordered to be changed in 1574, the former (No. 1138) design of the Holy Trinity being considered as idolatrous.

"s' ecclia collegiata sanct. trinitatis prope edibreg."—*From the original matrix of Copper, in the possession of J. Dimsdale, Esq., who purchased it at the sale of the library of Richard*

Gough, Esq. On the back of the matrix are engraved the letters M · R · P 1574 (*Dominus Magistri Robertus Pont*), which satisfactorily identifies it as the Seal of the College when Robert Pont was Provost. He was appointed to that *office in* 1571, *and was the last Provost. He was for some time one of the Lords of Session as well as minister of St. Cuthbert's Church, Edinburgh. Died in May* 1606.—*Hutton's "Seals,"* p. 152.

1140. ELCHO, Perthshire. A Nunnery of the Cistercian Order (see No. 1126), founded by David Lindsay of Glenesk.

Very imperfect, but a pretty design. Within a double niche, a nun, or female figure, praying to the Virgin and infant Jesus, standing. Inscription illegible, A.D. 1539.—*Don Charters.*

1141. ELCHO, EUFHEMIA LESLIE, Countess of Ross. Prioress of, resigned the Earldom of Ross in favour of the Earl of Buchan, A.D. 1382.

A fess chequé. In sinister chief three palets. Over all, on a bend, three buckles. The shield suspended on a crosier, and at the sides branches of foliage.

"S' EUFEAMIE LISCLE ABAS."—*Appended to Precept of Clare Constat in favour of John Swynton "de eodem" of the lands of Standertis, in Lothian,* 14th June 1394.—*Communicated by W. Fraser, Esq.*

1142. ELGIN, Dominicans or Friars-Preachers of (see No. 1111). Founded by King Alexander II., A.D. 1233. *Plate XII. fig.* 9.

A full-length front figure of a monk or pilgrim, with a staff in his left hand and a book in his right. The background ornamented with foliage.

"S' COMUNE [CAPLA LOCI] (?) DE ELGEIS ORDIS PREDICATORU."—"*Appended to a Charter by the Prior (John Spens) and Dominican Convent of Elgin to James Innes (second natural son to Alexander Innes of Innes) of the lands of Nether Meadonis and Brunactown,* 20th December 1546."—*Hutton's "Sigilla,"* p. 92.

1143. FERNE, Ross-shire. An Abbey of the Præmonstratenses (see No. 1129), founded by Ferquhard Earl of Ross, *circa* 1220.

A front figure of the Virgin sitting, with the infant Jesus in her arms. On the sinister side a pot with lilies.

"S' COMNE CAPITULI ET CONVENTUS MONASTERI DE FERNE."—*Appended to Charter by Thomas Ross, Commendator of Ferne and Provost of Tain, of the fourth part of the lands of Little Rayne, in the lordship of Ross and sheriffdom of Inverness, to Thomas Denune and Joneta Munro his spouse, at the Monastery of Ferne,* 3d August 1577.- *David Laing, Esq.*

1144. GLASGOW, UNIVERSITY of. Founded by W. Turnbull, Bishop of Glasgow, A.D. 1452.

Oval-shaped seal. Very rude work. A mace of office between a bird and a fish; above is a dexter hand holding an open book, over which is inscribed " via veritas via [vita ?]"

" s' commune universitatis glasgven."—Appended to a Tack of the Vicarage of Colmonel, Ayrshire, by the College of Glasgow to Thomas Kennedy of Kirkhill, 6th July 1688.—University of Glasgow.

1145. GLASGOW, JOHN, Archdeacon of.

A figure of a bishop, with a crosier in his left hand, his right extended, giving benediction to a monk kneeling before him, between them a fish with ring in its mouth.

" s' johis archid. glasgvens."—Brass matrix at Trinity College, Glenalmond, Perthshire.

1146. GLENLUCE, Galloway, St. Mary's.

Glenluce, or Vallis Lucis, was an Abbey of the Cistercians (see No. 1126), founded by Roland Lord of Galloway about A.D. 1190.

A full-length figure of the Virgin and infant Jesus, standing within a niche; at each side is a figure praying, and in the base a shield bearing a lion rampant, crowned, the feudal arms of Galloway.

" s' monasterii sant marie de vallis lucis."—Appended to " grant of lands in favour of the family of Kenmore, anno 1561."—Hutton's " Sigilla," p. 23.

1147. HADDINGTON, St. Mary's. A Convent of the Nuns of the Cistercian order (see No. 1126), founded by Ada Countess of Northumberland, Queen of David I. A.D. 1178, in honour of the Virgin Mary. Plate XII. fig. 1.

The Virgin, crowned, sitting with the infant Jesus on her knees, her feet resting on a crescent reversed; in her right hand she holds a cross, and in her left a branch of lilies.

" sigillum sancte marie de hapintune."—Appended to " Convention between J., the Prior, and Convent of St. Andrews and the Nuns of Haddington, 1245. In Brit. Museum."—Hutton's " Sigilla," p. 39.

1148. HOLYROOD, PATRICK, Abbot of the Monastery of the Canons Regular of St. Augustine.

This monastic order was instituted by St. Augustine, Bishop of Hippo, in Africa, in the fifth century. They were first introduced into Scotland early in the twelfth century, and met with greater encouragement than any other order, having no less than twenty-eight abbeys and monasteries, besides numerous cells and priories attached. Holyrood was founded by King David I., A.D. 1128.

This has evidently been a fine seal, though now much damaged. In the upper part of the seal, within a niche, is represented the Annunciation of the Virgin; in the centre, the crucifixion of the Saviour, and in base, a bishop praying, his mitre at the side.

" s' patricius sce crucis de edinburgh."—*Appended to an Instrument relating to the churches and chapels of Coldingham, A.D. 1455.—Dean and Chapter of Durham.*

1149. HOLYROOD, ARCHIBALD CRAWFORD, Abbot of.

A shield suspended on a pastoral staff, and bearing a fess chequé.

" s' archibaldi . . . ."—*Appended to Truce between England and Scotland, 12th December 1465.—H.M. Record Office.*

1150. HOLYWOOD (Sacri Nemoris) Galloway. A Monastery of the Præmonstratenses (see No. 1131), founded early in the thirteenth century. *Plate XV. fig. 2.*

A bird sitting on the branch of a tree; in the lower part are two estoiles.

" s' com abbis et conventi sac nemonis."—*Appended to a Lease by Thomas (Campbell), Abbot of Holywood, dated 15th November 1557.—Hutton's " Sigilla," p. 114.*

1151. ICOLMKIL, Isle of Hy or Iona. A Monastery of the Cluniacenses. (See No. 1127.)

It is much to be regretted that the only seals of this celebrated monastery that have yet been met with are so much injured that an accurate description can scarcely be given. This appears to be a figure of St. Columba and a monk kneeling at each side, standing in the centre of an archway supported by towers. From the style of the work and form of letters in the small part of the inscription that remains, it may be ascribed to the thirteenth or early part of the fourteenth century.

" sigillu. canonacorum yensis mo . . . i sci cluc . . . ."—*From General Hutton's Collection.*

1152. JEDBURGH, CHAPTER of. An Abbey of the Canons Regular of St. Augustine (see No. 1148), founded by King David I. *Plate* XIII. *fig.* 1.

Very fine design, but in bad preservation. The Father placing a crown on the head of the Virgin, sitting beside him.

" SIGILLU[M CA]PITULI DE JEDEWERTHE."

1153. COUNTER SEAL of the last. *Plate* XIII. *fig.* 2.

The design, within a highly decorated niche, represents the Salutation of the Virgin. " MATER CASO . . . SERVIS . . . ANIA."—*Appended to Sasine of the lands of Cossins, A.D.* 1534.—*Glammis Charters.*

1154. JEDBURGH, ANDREW HOME, COMMENDATOR of.

The Virgin and infant Jesus standing within a Gothic niche. In base of seal is a shield, quarterly: first and fourth, a lion rampant, for Home; second and third, three papingoes, for Pepdie of Dunglas.

" S' ANDREE COMMENDATARE MONASTERII DE JEDBURGH."—*From a bad mateir in Dr. Rawlinson's Collection.*—*Bodleian Library, Oxford.*

1155. KELSO, ROBERT, ABBOT of. A Monastery of the Tyronenses, founded by King David I., A.D. 1128. It was founded at Selkirk some years previously, but removed to Kelso. The Tyronenses were an order instituted early in the twelfth century, and followed the rule of St. Benedict. (*See* No. 1120.)

A full-length front figure of an abbot, mitred and robed, holding in his right hand a crosier and in his left a book. At the dexter side a rose, and on the sinister an estoile and crescent.

" SIGILLUM ROBERTI ABBATIS DE CALCHOU."—*Dean and Chapter of Durham.*

1156. KELSO, CHAPTER of.

A richly designed seal, but now much injured. The Virgin sitting with the infant Jesus, within a niche.

" S' COM[MU]NE CAPITULI DE CAL[CHOU]."

1157. COUNTER SEAL of the last.

A highly-ornamented triple niche. In the centre one the figure of a bishop, in his right hand a crosier, his left elevated. In each of the side niches a monk holding an altar candle. In base three leaves. Legend not read.—*Appended to a Deed by the Abbot (William) and Convent of Kelso, dated* A.D. 1350.—*Hutton's* " *Seals,*" p. 32.

1158. KINLOSS, Moray, THOMAS, Abbot of. An Abbey of the Cistercian Order
(see No. 1126), founded by King David I., A.D. 1150.

A full-length front figure of a monk holding a crosier, standing within a niche. The
background ornamented with foliage. In base a shield, bearing apparently
a tree growing from a mount.

"SIGILLUM THOME ABBATIS DE KINLOS."—*Appended to a Precept of Sasine dated
8th April 1525.—Hutton's "Seals," p. 113.*

1159. LINDORES, Fife, Abbey of. An Abbey of the Tyronenses (see No. 1155), founded
by David Earl of Huntingdon about A.D. 1178. *Plate XIII. fig. 4.*

A fine seal, in bad condition, with the design of the martyrdom of St. Andrew.
On the sinister side is a man on a ladder tying the arm of the saint to the
cross, and on the dexter a group of ten monks singing praises, while below is
a monk praying between a mullet on the dexter and a crescent on the sinister.

"SIGIL. CONVENT SANC MARIE ET S. ANDRE . . . BAT . . . (?)."

1160. COUNTER SEAL of the last. *Plate XIII. fig. 5.*

The Virgin, with the infant Jesus crowned with the nimbus on her knee, sitting
beneath a canopy. On the dexter side is a monk praying or singing; a scroll
before his mouth, inscribed with the words "AVE MARIA." On the sinister
side is a group of four monks, also chanting or praying, and a scroll, with the
words "SALVE SCA I . . . ."

"S' CAPITULI ECCLIE [SCE MARIE ET SCI ANDREE] DE LUNDORES."—*Appended to a
Charter by the Abbot (John) and Convent of Lindores to Patrick Leslie (son of
Alexander Leslie of Wardross ?) of the lands of Flanders, 20th March 1554.
—Hutton's "Sigilla," p. 82.*

1161. LINLITHGOW, PRESBYTERY of.

An open book, on which is inscribed "VERBUM AUTEM DEI NOSTRI STABIT IN ETER-
NUM. ISA. 40," (being the last clause of the eighth verse of that chapter), and
below it the date "1583."

"SIGILLUM PRESBETERII LINLICHQUO."—*From the original matrix in possession of the
Presbytery of Linlithgow.*

1162. MANUEL, Linlithgowshire. A Convent of Cistercian Nuns (see No. 1126),
founded in honour of the Virgin Mary by Malcolm IV., A.D. 1156.

A full-length front figure of the Virgin and child; background foliated. Rather
rude work.

"S' COMMUNE MONASTERY DE [MANU]EL."—"*Appended to a Charter by the Prior*

(James Hopper) and Convent of the Carmelites of Linlithgow, 11th January 1559."—Perhaps they had lost their own seal.—Hutton's "Seals," p. 125.

1163. MELROS, Roxburghshire. An Abbey of the Cistercian order (see No. 1126), founded by King David I., A.D. 1136, and dedicated to the Virgin Mary.

A full-length front figure of a monk, his right hand raised, giving the benediction, his left holding the crosier.

"SIGILLUM ABBATIS DE MELROS."—*Dean and Chapter of Durham.*

1164. MELROS, Abbot of.

An arm vested, issuing from the sinister side, holding a crosier ; the background diapered of a lozenge pattern.

"MANUS ABBATIS DE MELROS."—*Appended to an Acquittance relative to Tithes at "Ersilton," without date.—Dean and Chapter of Durham.*

1165. MELROS, ARNOLD, Abbot of.

A full-length front figure of a monk with a cross in his right hand, and a crosier in his left.

"SIGILLUM ABBATIS DE MELROS."—*Appended to Transcript of a Charter by Robert, Bishop of St. Andrews, with consent of David, King of Scotland, regarding the liberties of the Church of Coldingham and other churches and chapels, dated A.D. 1127.—Dean and Chapter of Durham.*

1166. MELROS, ANDREW HUNLAC, Abbot of.

A figure of a monk in front, crowned with the nimbus, holding a crosier in his right hand, and a book in his left, standing in a niche.

"SIGILLUM FRATRIS ANDREE HUNLAC X GRA ABBATIS DE MELROS."—*Appended to an Instrument regarding the union of the Churches of Abbcombus and others, A.D. 1466.—Dean and Chapter of Durham.*

1167. MORAY, Chapter of.

An exceedingly rude figure of a saint sitting.

"SIGILLUM CAPITULI MORAVIENSIS ECCLESIE, 31 MARCH 1585."—*From the original brass matrix at Trinity College, Glenalmond, Perthshire.*

1168. NEWBOTTLE, Mid-Lothian, THOMAS, Abbot of. An Abbey of the Cistercians (see No. 1126), founded by King David I., A.D. 1140.

Within a Gothic niche, figures of a monk holding a crosier in his right hand, and in his left a book.

"SIGILLUM DOMPNI (?) THOME ABBATIS DE NEUBOTLE."—*Appended to an Instrument dated A.D. 1145.—Dean and Chapter of Durham.*

1169. NEWBOTTLE, Chapter of the Monastery of.

> A figure of the Virgin, with the infant Jesus in her arms, standing within a Gothic niche, on the dexter of which is a shield bearing the arms of Scotland, and on the sinister side a shield charged with a fess and five piles.
>
> "S' COMMUNE CAPITULI MARIE DE NEWBOTTLE."—*Appended to Charter of Poverhow, etc., to Andrew Murray of Blackbarony, 12th March 1552.—Blackbarony Charters.*

1170. NEWBOTTLE, MARK KERR, Commendator of, ancestor of the Marquis of Lothian.

> Full-length figure of the Virgin and infant Jesus within a Gothic niche ; in lower part of the seal is a shield bearing, on a chevron, three mullets ; in base a unicorn's head couped.
>
> ". . . . MARCUS COMMENDATARIUS DE SEC."—*Appended to Charter of Lands in Monkland, Lanark, to Robert Boyd, son of Lord Boyd, 17th September 1563.—Errol Charters.*

1171. PEEBLES, The Holy Cross of. A Hospital or Ministry of the Trinity or Redfriars (called also St. Mathurines), an order established in the twelfth century for the redemption of Christian prisoners. They had several houses in this country ; Peebles was founded by King Alexander III., A.D. 1257.

> A cross in front of a chapel. The impression is very imperfect, and the legend quite illegible.—*Appended to " an Obligation by the Minister (John Madowr) of the Cross Kirk of Peebles to say a daily mass at the Black-rude altar of the Cross Kirk for the souls of an honourable man, Thomas le Hay, vice deputato de Peblis, and Christiane his spouse, 18th December 1484." " Copied from the original, belonging to the Duke of Queensberry, 1796."* Hutton's "Sigilla," p. 183.

1172. PEEBLES, ROBERT STUART, Minister of the Cross Kirk of.

> The same design as the preceding, except that in this instance the cross rests on five steps.
>
> " SIGILLUM ROBERTI STUART MINISTERI DE PEBLES."—*Appended to a Tack or Assignation, dated A.D. 1597.—Communicated by G. Logan, Esq., Teind Office.*
>
> This document is signed by King James VI., and " in syne of his Majesty's consent and assent as patron, his hieness previe seill, togidder with the common seil of the charter of the Cross Kirk, etc. etc., is appendit."
>
> It is worthy of remark that this " previe seill " of James VI. is the one used after his accession to the English Crown. The shield bears, first and fourth, Scotland ; second, France and England, quarterly ; third, Ireland. How it could

have been appended to a document executed in 1597 can only be explained by supposing the deed had not passed the Privy Seal till some years subsequently.

1173. PERTH, "VALLIS VIRTUTIS," THE CHARTER-HOUSE. A Monastery of the Carthusians, an order instituted in the eleventh century. They followed the rule of St. Benedict, with some additional severities. This establishment at Perth—the only one they had in Scotland—was founded by King James I., A.D. 1429. It is said to have been a fine structure. The seal of the House is a beautiful example.

Beneath a Gothic canopy, the crowning of the Virgin; in the lower part of the seal is a monk kneeling on a cushion, his arms uplifted, and his head thrown back; before him is a crown, and on a scroll is inscribed " RADIATE MEA."

" S' DOMUS VALLIS VIRTUTIS ORD CARTUSIE IN SCOTIA."—*Appended to a Charter by David Symonton, with consent of the Chapter, to Donald Macande, of a croft of land in Killin, 20th November 1488.—Breadalbane Charters.*

1174. PERTH, FRIARS-PREACHERS OF. A Monastery of the Dominicans (see No. 1141), founded by Alexander II., A.D. 1231.

A rudely executed design of the Virgin and Child within a niche; in the lower part a monk praying.

" S' OFFICLL. POTI ORD. PRICARII DE PTH."—*Brass matrix in the Museum of the Society of Antiquaries of Scotland.*

1175. QUEENSFERRY (SOUTH), CARMELITES OF. A Monastery of the Carmelites (see No. 1108), founded by the " Laird of Dundas," A.D. 1330.

A figure of the Virgin with the infant Jesus in her arms, standing on a crescent, surrounded by an aureole.

" S' COE LOCI FRATU. CARMELITARU. PORTUS REGINE."—" *Appended to a Charter by Thomas Young, Prior of the Carmelites of Queensferry, to George Dundas of Dundas, of a tenement in Queensferry, dated 1564."—Hutton's " Seals," p. 51.*

1176. RESTALRIG. A Collegiate Church near Edinburgh, founded by King James II.
A full-length figure of the Virgin with the infant Jesus in her arms; the background foliated.

" s' COMMUNE RESTALRIC." (?)—*Appended to a Mandate by W. Gibson, Dean of Restalrig, and others, Commissioners appointed by the Pope (Paul III.) for confirming a feu charter by the Prior and Convent of Pittenweem, to W. Cockburn of Cockburn, of the lands of Mayscheil, in Berwickshire,* A.D. 1535. —*David Laing, Esq.*

1177. ROSS, CHAPTER OF ST. PETER'S.
Large, oval shape. A copy of the fine old seal of the Chapter (No. 1102 " *Descriptive Catalogue of Ancient Scottish Seals*"), but very inferior in style of art. Figures of St. Peter and St. Boniface; the background ornamented with foliage.

" s' CAPITULI SCOR. PETRI ET BONEFACIE ROSMARRIS."—*Appended to the same Instrument as No.* 847.

1178. ST. ANDREWS, CHAPTER OF ST. MARY'S, KIRKHEUGH. This was a Collegiate Church, formerly a church of the Culdees; very little is known of its early history. It was the first Chapel-Royal of the Scottish Kings. The foundations have been recently discovered on an eminence overlooking the harbour of St. Andrews, and are now carefully preserved. The seal has rather an imposing appearance, but the art is rude, and obviously that of the sixteenth century.

A triple niche, in the centre one the Virgin, crowned, sitting with the infant Jesus in her arms; in each of the side niches is a monk kneeling at prayer.

" s' CAPITULI ECCE SCE MARIE CAPELLE DOM. REGIS SCOTTORUM."

1179. COUNTER SEAL OF THE LAST.
The king seated, with both arms extended, his right hand holding a sword or sceptre, his left the orb; the background seems to have been diapered with fleurs-de-lis, but the impressions of both these seals are imperfect.

The inscription is the same as on the last.—*Appended to a Charter by James Learmonth, Provost of St. Mary's, Kirkheugh, near St. Andrews, to Andrew Ayton of Kinaldy, 1st February* 1575.—*Kinaldy Papers.—Communicated by David Smith, Esq., St. Andrews.*

1180. ST. LEONARDS, NEAR PERTH. A Cistercian Priory (*see* No. 1126), founded at an early period, but suppressed by King James I. and annexed to the Charter-house.

A pretty seal, having the design of the Virgin (half-length) with the infant Jesus in her arms, within a niche. In base a monk or nun at prayer.

" S' SORORIS STEPHANIE DCE DE KELMARO."—*Appended to Homage Deed* A.D. 1296. —*H.M. Record Office.*

1181. ST. NINIAN'S HOSPITAL OR CHAPEL.

An interesting seal of an institution, early established " at Leyth Wynd fute," but of which very little is known. Unfortunately it is but a fragment; and from the style of work is probably not much earlier than the date of the instrument to which it is appended. The design represents a building, doubtless meant for the hospital; and in front is an almsman soliciting relief. All that remains of the inscription is the following:—" . . . QUI LARGITUR AV . . . ," probably part of Proverbs xix. 17.—*Appended to an Instrument dated* 16th June 1605. —*Edinburgh Charters.*

1182. SCONE, PERTHSHIRE. An Abbey of the Canons Regular of St. Augustine (*see* No. 1148), founded by King Alexander I., A.D. 1114, in honour of the Holy Trinity, and St. Michael, the Archbishop.

A fine seal, though damaged, representing a church with a lofty central spire, and lower ones at each end. On the roof are two figures sitting, one on each side of the central spire.

" SIGIL. ECLESIE SANCTE TRINITATIS DE SCONA."—*Appended to* " *an Indenture between the Abbot and Convent of Scone and Davd. de Hayo, anno* 1237," *at Slains Castle.*—*Hutton's* " *Sigilla,*" p. 67, *and* " *Seals,*" p. 137.

1183. SWEETHEART. DULCIS-CORDIA, ALSO CALLED NEW ABBEY. IN GALLOWAY. An Abbey of the Cistercian Order (*see* No. 1126), founded by Lady Devorgilla Balliol in the thirteenth century, who here deposited the heart of her husband, whence the name of the abbey.

The Virgin with the infant Jesus, sitting in a Gothic niche, both crowned with arched crowns. Rather rude work.

" SIGILLUM COMMUNE (CONV) DULCIS CORDIA."—*Appended to a Tack of the Teinds of Locghil, in Kirkcudbright, to John Browne of Locghil, for nineteen years,* 4th October 1594.—*David Laing, Esq.*

1184. SCOTLAND, CARMELITES of. (See No. 1108.) Plate X. fig. 2.

A round seal, with a well-executed design of St. Andrew on the cross, crowned with the nimbus. At each side is a branch of foliage ; a crescent and a mullet above, and in the lower part a monk kneeling at prayer, within a niche.

" s' comune frm carmelitar scocie."—*Appended to Charter by " Frere Robert provincile generale off ye Order of Carmelites within ye Realme of Scotland," 30th October 1492.—Dundas Charters.*

1185. SOUDAN, ADAM, Rector of.

A full-length figure of St. Catherine, holding in her left hand the wheel, the instrument of her martyrdom ; her right supports a sword, the point resting on it. At her left side is a monk praying to her.

" s' ade rectore de soudan."—*This pretty seal is appended to a Homage Deed. A.D. 1296.—H.M. Record Office.*

1186. WHITHORN (Candida Casa), Galloway   A Priory of the Præmonstratenses. (See No. 1129.) Plate XII. fig. 7.

A building, probably intended to represent the church.

" sigillum prioris et capituli candide casa."   Detached Seal.—*From General Hutton's Collection.*

1187. WHITHORN (Candida Casa). Chapter of Galloway.

An oval-shaped seal. A church having a door and three windows. From the centre of the roof rises a square tower, terminating in four arches and decorated with fanes. In the base of seal are sprigs of laurel, etc.

" sigillum capituli candide cassa episcopi, 1613."—*Appended to the same Instrument as No. 1094.*

MISCELLANEOUS OFFICIAL SEALS.

1188. JUSTICIARY COURT for the Lothians. Reign of King David II.

A shield. The initial D. within the royal tressure. Above the shield a crown of five points (fleurs-de-lis) and at the sides an ornament of foliage. Very prettily executed.

" S' OFFICII JUSTICIARIE LAUDONIE."—*Appended to a Declaration made by the Prior of Coldingham before a Justiciary Court held at Melrose, 6th June 1366, presided over by Robert Erskine, Lord Justice.—Dean and Chapter of Durham.*

1189. JUSTICIARY COURT for the District North of the Forth.

A well-executed seal, with the royal arms of Scotland. Crown above, and foliage as in the last.

" S' OFFI. JUSTII DE REGM P. BOR. ACQUA DE FORT."—*Affixed on paper, being a Remission to David Pitcarne of Trotter, by Sir W. Scott of Balweary, Lord Justice, 12th April 1522, signed by " Master James Wischart of Pitarro, Clerk of Justiciarie."—David Laing, Esq.*

1190. COUNCIL OF SCOTLAND. Time of the Commonwealth.

An ornamented shield, surrounded by laurel branches, charged with a saltire, surmounted with an escutcheon bearing the Scottish lion, but without the royal tressure.

" SIGILLUM CONCILII SCOTIE."—*From a Letter to the Commissioners for Assessment in the shire of Orkney and Shetland, dated 11th May 1658 ; signed by General Monk.—David Laing, Esq.*

1191. FORFAR, SHERIFF of (David Earl Crawford, Duke of Montrose).

A fess, counter compony, between two crowns in chief, and a mace in base.

" S' OFFICII VIC COMIT DE FORFAR."—*Appended to the Retour of Walter Lindsay, as heir to his deceased brother William Lindsay, in the lands of Loquharry, in the barony of Ferne, co. Forfar, 29th January 1469. The retour is made in the court of " nobile vir Alexr. Lindsay. vice, com Depute de Forfar."—Crawford and Balcarras Charters.*

1192. INVERNESS, SHERIFF of (George, fifth Earl of Huntly).

A shield, between the initials G · H., bearing the coats of Gordon, Badenoch, Seton, and Fraser, quarterly.

" S' OFFICII VICE COMITATISS DE INVERNIS."—*Appended to Transumpt of a Charter to William Dallas, 30th July 1590.—Cawdor Charters.*

1193. LYON KING OF ARMS, Office of.

A lion sejant, affronté, holding in the dexter paw a thistle, slipped, and in the sinister a shield uncharged. On a chief, a saltire. At each side of the shield is a branch of laurel.

" SIGILLUM OFFICII LEONIS REGIS ARMORUM. 1673."—*From the silver matrix in the Lyon Office.*

1194. HERIOT'S HOSPITAL, Edinburgh.

A view of the north front of the building, erected in the year 1650. At each side of the spire is an escutcheon, that on the dexter bearing a mullet, and on a chief three cinquefoils. The shield on sinister side of the spire bears the initials G · H., between a cinquefoil in chief and a thistle leaved and slipped.

" SIGILLUM HOSPITALIS GEORGII HERIOT."—*From the silver matrix made in 1649, in the custody of Isaac Bayley, Esq., Clerk of the Hospital.*

1195. HERIOT, GEORGE. The celebrated goldsmith to King James VI., and founder of the above Hospital.

A small signet, with the device of a dexter hand issuing from clouds, grasping a goldsmith's hammer above an anvil, on which is inscribed, " FERENDO FERIO."

This seal was used by Dr. Balcanquall, Dean of Rochester, the executor of George Heriot, subsequently to A.D. 1623, the year of his death; but the device and motto seem to offer at least presumptive proof that it was the seal of George Heriot, who probably had bequeathed it to Dr. Balcanquall, as it occurs in sealing the original statutes of the Hospital, 13th July 1627.—*Governors of Heriot's Hospital.*

1196. DUNDEE, INCORPORATION OF HAMMERMEN of.

A figure of St. Eloi, in episcopal vestments, holding a hammer in his right hand and a crosier in his left, standing within a niche: at each side is a vase or pot of lilies, and beneath is a shield bearing a hammer in pale and a crown of three points in chief.

" S' CE ARTIS MALLIATO SCI ELEGI DE DUNDE."—The design is well executed, and the original brass matrix is in excellent preservation, in the Advocates' Library, Edinburgh. It is probably the work of the fifteenth century.

BURGH SEALS.

1197. ABERDEEN. The capital of the county, and a very ancient royal burgh. The earliest charter extant in the Archives of Aberdeen is by William the Lion, A.D. 1179. *Plate* XIV. *fig.* 6.

A figure of a bishop—St. Nicolas,—mitred and robed, bestowing the benediction. On the dexter side a crescent, and on the sinister a mullet of six points.

"SIGNUM BEATI NICOLAI ABERDONENSIS."

1198. COUNTER SEAL of the last. *Plate* XV. *fig.* 8.

A church, or other edifice, with a large central spire and two smaller ones, each terminating in a cross.

"SIGILLUM DE COMMUNI ABERDONENSIS."—*Appended to Procuration by the Burghs of Scotland for the ransom of King David II., 26th September* 1357.—*H.M. Record Office.*

The style and workmanship of this seal conclusively prove it to have been executed much earlier than the date of this instrument. It is conjectured by Mr. Astle, in his Account of the Burghs, etc., of Scotland, that the triple-coned building represents the mausoleum or shrine of the patron saint, and it is supposed by others that these spires have formed the type of the three triple-towered castles which for four centuries at least have figured on the shield of Aberdeen. This is more probable than the popular tradition of their origin, which, as usual in similar cases, rests on no authority.

1199. AYR. The capital of the county, created a royal burgh in 1202. *Plate* XIV. *fig.* 4.

A castle with three embattled towers, having spires terminating in balls. At the dexter side is the " Agnus Dei," and on the sinister the head of John the Baptist in a chalice, the whole enclosed within an angular moulding, perhaps meant to represent the fosse surrounding the castle.

" S' COMMUNE BURGI DE ARE."—*Hutton's " Seals."*

1200. BANFF. The capital of the county. Erected into a royal burgh 1372.

An oval-shaped seal. A boar passant.

" S' COMMUNE DE BANFS." Rudely executed.—"*Appended to a Charter,* A.D. 1408. *See Forglen Papers."—Hutton's Collection.*

1201. BANFF.

The same design, but of inferior workmanship.

"INSIGNIA URBIS BANFIENSIS."—*Matrix.—Office of Town-Clerk.*

1202. BERWICK.　The capital of the county; a burgh at an early period.

A very fine seal, but unfortunately rather injured.　A bear chained to a tree, in the branches of which are two birds (parrots?) surrounded by a double tressure of fleur-de-lis.

"S' . . . T . . . VILLE BERWIC . . . EDAM."

1203. COUNTER SEAL OF THE LAST.

A representation of the Holy Trinity.　The Father sitting, supporting, between his knees, the Son on the cross, and the Holy Spirit descending on his head.

"BENED . . . SANCTA TRIN . . . ."—*Appended to a Document dated February* 1330.—*Dean and Chapter of Durham.*

1204. CRAIL.　A royal burgh in Fife, which obtained its privileges by a charter from King Robert the Bruce.

In bad condition, but has been a good seal.　A galley with rowers.

1205. COUNTER SEAL OF THE LAST.

The Virgin and infant Jesus, with angels worshipping.　The inscription on both these impressions is quite lost.—*Appended to the same Instrument as No.* 1196.

1206. CRAIL.

A small seal of the same burgh.　A galley, with sails furled.　Stars and crescent above.

"SIGILLUM COMMUNE BURGI DE KARALE."—*Original matrix.—Town-Clerk of Crail.*

1207. CULLEN.　A royal burgh in Banffshire.

Very rude workmanship.　The Virgin and infant Jesus in her arms, seated on a kind of throne; below is a figure of a dog.

"SIGILLUM URBIS DE CULLEN."—*From the matrix in the Office of the Town-Clerk.*

1208. CULROSS. PERTHSHIRE.　Created a royal burgh by King James VI., A.D. 1588.

A small seal, with the representation of the parish church of Culross.

"THE ROYAL BURGH OF CULROS."—*From the original matrix in the Office of the Town-Clerk of Culross.　A larger seal of this burgh, also preserved in the same office, having a similar design, is described under No.* 1151 *of* "Descriptive Catalogue of Ancient Scottish Seals."

1209. DINGWALL, Ross-shire. The capital of the county. Erected a royal burgh by King Alexander II. *Plate XV. fig.* 1.

    A star-fish or estoile. Between the points are two lozenges, a heart and two mullets.

    " SIGILLUM COMUNE BURGI DE DINGWALL."—The present seal of the burgh. Not earlier work than the last century, but a copy apparently of the older seal. - *E. Colquhoun, Esq.*

1210. DUNBAR, Haddington. Created a royal burgh by King David II.

    An elephant, with a castle on its back.

    " SIGILL. [COMU]NE BURGI DUN."—*Appended to same Instrument as No.* 1196.

1211. DUNDEE, Forfarshire. Created a royal burgh by William the Lion. *Plate XIV. fig.* 2.

    The Virgin and infant Jesus, seated. At each side is an angel waving the thurible.

    " SIGILLUM COMMUNE VILLE DE DUNDEE."

1212. COUNTER SEAL. *Plate XIV. fig.* 1.

    A figure of St. Clement, sitting, mitred, and crowned with a nimbus, holding in his right hand an anchor; a monk kneeling before him. All that can be read of the inscription is " SIGNUM SANCTI [CLEM]ENTIS DE DUNDEE." - *Appended to the same Instrument as the last.*

1213. DUNFERMLINE, Fife. Constituted a royal burgh by King James VI.

    A full-length figure of St. Margaret, crowned, holding a sceptre in her right hand, standing within a Gothic niche, at each side of which is an altar-candlestick.

    " S' MARGARETA REGINA SCOTORUM.'

1214. COUNTER SEAL of the last.

    A tower or fort, supported by two lions. Above it is inscribed, " ESTO RUPES INACCESSA." Surrounding inscription, " SIGILLUM CIVITATIS FERMILODUM."— *From an impression in lead, in possession of Dr. E. Henderson, Inverkeith.*

1215. EDINBURGH. The capital of Scotland, and a royal burgh as early as A.D. 1128.

    A castle, single towered, gates closed. Foliage in the background.

    " SIGILLUM COMETTI (?) DE EDINBURGH.'

1216. COUNTER SEAL of the last.

    The arms of Scotland. The shield surrounded by foliage.

      " JACOBUS DEI GRACIA REX SCOTTORUM."—*From the Collection of Casts taken by the late Mr. Doubleday.*

1217. FALKLAND, Fife. Erected a royal burgh A.D. 1458.

    A stag lying in front of a tree and looking upwards.

      " DISCITE JUSTITIAM MONITI NON TEMNERE CHRISTUM."—*Matrix in the Office of the Town-Clerk of Falkland.*

1218. FORRES, Morayshire. An early royal burgh.

    A full-length figure of St. Laurence, crowned with a nimbus, holding a book in his right hand, his left resting on the gridiron, the instrument of his martyrdom. In the field is a crescent, a star, and two branches of foliage.

      " SIGILLUM COMMUNE BURGI DE FORES."—*From the original brass matrix in the Office of the Town-Clerk of Forres.*

    This is an excellent example of fifteenth century work. It has been broken in two, but is now soldered together, and is still used by the burgh.

1219. HADDINGTON. The capital of the county, and a royal burgh. *Plate* XV. *fig.* 6.

    A figure of the King (David I.), crowned, sitting on a throne, his right hand resting on a shield bearing the arms of Scotland, his left holding a sceptre, terminating in a fleur-de-lis.

      " DAVID DEI GRATIA REX SCOTTORUM."—At the commencement of the inscription is a singular blank space, which can only be accounted for by supposing that the engraver had cut a wrong word, probably " SIGILLUM," and rather than sacrifice the whole of the work has clumsily cut it out, thus leaving a deep blank space, very injurious to the beauty of the design.

1220. COUNTER SEAL of the last. *Plate* XV. *fig.* 7.

    A goat reared on its hind legs, browsing on an apple-tree (?). The background diapered of a lozenge pattern.

      " SIGILLUM COMMUN. BURGI DE HADINGTON."—*Appended to Extract of Process in the Burgh Court of Haddington,* 12th October 1518.

    The style and design of this fine seal leave no doubt of its being the work of the thirteenth, or early part of the fourteenth century.

1222. HADDINGTON.

A goat passant.

" s' . . . ARIUM BURGI DE HADINGTOUN."

The reverse of this is the same as No. 1219, differing only in having two goats browsing on the tree. A careful and minute examination leads to the conclusion that both impressions are from the same die, and the additional goat engraved on it between A.D. 1518, the date of the former, and 1578, the date of this. No reason can now be given for the addition, except, perhaps, to give symmetry to the design.—*Appended to a Document relating to a debt due to Patrick Brown of Colston, from a tenement in Haddington, 18th November 1578.—Colston Charters.*

1223. INVERKEITHING, FIFE. An early royal burgh.

Much damaged, and rudely executed. A figure of a bishop sitting, and in his right hand holding the model of a church.

1224. COUNTER SEAL.

A galley on the water.

" s' COMMUNE . . . INVERKETHYN."—*Appended to the same Instrument as No. 1197.*

1225. INVERNESS. The capital of the county, and a royal burgh. Certainly one of the earliest burghs possessing such privileges in Scotland.

A camel passant. Above the shield, "INVERNESS." Inscription around, "FIDELITAS ET CONCORDIA."—*Brass matrix in the office of the Town Clerk of Inverness.*

This seal is probably not earlier than the last century. The armorial ensigns differ materially from those displayed in the Town-Hall of Inverness, where they are well executed in stone, and ornament the wall of the staircase in that edifice. There they appear as the Saviour on the cross, with a camel as the dexter supporter of the shield, and an elephant for the sinister. Crest, a cornucopia, and on a ribbon above, " CONCORDIA ET FIDELITAS."

The design here is but the reverse of the early common seal of the burgh (see No. 1167, " *Descriptive Catalogue of Ancient Scottish Seals*") placed on a shield, with the addition of crest and supporters, these latter having been assumed as representative of the extensive trade once carried on between the port of Inverness and the east.

The discrepancies so obvious in the armorial ensigns of the burgh are much to be regretted, and still more, perhaps, that the good burgh has not obtained legal authority for the use of any.

2 E

1226. IRVINE, AYRSHIRE. A royal burgh previous to the reign of Alexander II., A.D. 1240. The Virgin sitting on a chair with the infant Jesus in her arms, both crowned with the nimbus, within a kind of Gothic niche.

"S' COMUNE DI RGI DE IRVINE."

1227. COUNTER SEAL OF THE LAST.

A lion sejant, affronté, holding in the right paw a sword, and in the left a sceptre, erect, both paws extended. Foliage in the lower part of the seal. Inscription the same as in the last.—*Matrix preserved in the office of the Town-Clerk of Irvine.—Communicated by E. Colquhoun, Esq.*

1228. IRVINE.

A smaller seal of the same burgh as the last. A lion sejant, gardant, crowned, between two trees.

"S' COMMUNE BURGI DE IRVINE."

Both these seals of Irvine are probably not earlier than the seventeenth century, and of very poor work.—*Communicated by E. Colquhoun, Esq.*

1229. JEDBURGH, ROXBURGHSHIRE. The county town of Roxburghshire, and an ancient royal burgh.

Oval shape. The Virgin sitting in a chair, opposite another figure, also in a chair, who appears to be addressing her, and offering a book; crescents, stars, and points ornament the background, all surrounded by rude tracery.

"S' COMMUNITATIS DE JEDDEWORTHE." Detached Seal.—*H.M. Record Office.*

1230. JEDBURGH.

The Virgin and infant Jesus within a niche, ornamented at the sides with foliage, and a scroll, on which is inscribed the names "MARIA." "JESUS."

"SIGILLUM COMMUNE BURGI DE JEDBURGH."—*From General Hutton's Collection.*

1231. JEDBURGH.

A shield, charged with an armed knight at full gallop. On a scroll above the shield, "STERNUS ET PROSPERE," and on one below, "SIGILLUM BURGI DE JEDBURGH."—*From General Hutton's Collection.*

1232. JEDBURGH.

On a shield, a unicorn passant.

"S' COMUNITATIS DE JEDBURGH."—*From General Hutton's Collection.*

1233. KILRENNY, Fife. Created a royal burgh by King James VI.

A fishing-boat, with four men rowing and one steering; a hook hanging over the side.

" SEMPER TIBI PENIAT HAMUS KILRENNY."—*Matrix in the office of the Town-Clerk of Kilrenny.—Communicated by E. Colquhoun, Esq.*

1234. KINGHORN, Fife. An early royal burgh, said to have been so created by King David I.

A castle; a mullet or cinquefoil at each side.

" s' E (?) COMUNE BURGI DE KINGORNE"

1235. COUNTER SEAL of the last.

A full-length figure of St. Leonard, with a crosier in his left hand, his right hand raised; foliage ornaments the background.

" SANCTUS LEONARDUS DE KINGORN.". These are executed in a very rude style of art, and are probably the work of the sixteenth century.—*Matrices preserved in the office of the Town-Clerk of Kinghorn.*

1236. KINTORE, Aberdeenshire. A royal burgh as early as the twelfth century. *Plate XV. fig. 3.*

Oval-shaped seal. A branch of foliage well designed.

" s' COMMUNE DE KINTOR."—*Matrix in the office of the Town-Clerk of Kintore.*

1237. KIRKCUDBRIGHT. Capital of the county, erected a royal burgh in 1455 by King James II. *Plate XIV. fig. 5.*

A good example of fifteenth-century work. St. Cuthbert, sitting in a boat, with the head of St. Oswald on his knees. The design is on a shield, at the sides and top of which is foliage.

" SIGILLUM COMUNE DE KIRKCUBRYU." St. Oswald, a celebrated Saxon prince of the seventh century, who embraced Christianity, and was killed in battle with Penda, King of Mercia, who cut off his head and fixed it on a stake, where it remained till rescued by Oswy, King of Northumberland, and placed in the tomb of St. Cuthbert.—*Matrix in the office of the Town-Clerk of Kirkcudbright.—Communicated by E. Colquhoun, Esq.*

1238. KIRKWALL. The capital of the Orkney Islands, erected a royal burgh by King James III.

A three-masted galley on the waters, sails furled.

" SIGILLUM COMMONE CIVITATIS KIRKUALENSIS, 1675."—*Matrix in the office of the Town-Clerk of Kirkwall.—Communicated by E. Colquhoun, Esq.*

1239. LANARK. The capital of the Upper Ward of Lanarkshire, and a royal burgh in the reign of Malcolm IV.

An eagle displayed, with two heads, not on a shield, between two lions rampant, in the upper part, and two fishes (salmon?) in the lower; the background ornamented with annulets.

"SIGILLUM COMMUNE BURGI DE LANARCK."—*Communicated by E. Colquhoun, Esq.*

1240. LANARK.

The same figures as the preceding, but here placed heraldically on a shield, the two lions in chief being passant counter-passant. Inscription as before.—*Communicated by E. Colquhoun, Esq.*

1241. LOCHMABEN, DUMFRIESSHIRE. Created a royal burgh by King Robert the Bruce.

A full length front figure of a female with long flowing hair, holding in her left hand a casket or reliquary; the background diapered with a lozenge pattern.

"S' COMUNI VILLE DE ROI DE LOHMABEN."—*Hutton's "Seals," p. 90.*

1242. MAYBOLE, AYRSHIRE. Created a burgh of barony in 1516.

A building, probably intended for the church or town-hall of Maybole.

"THE SEAL OF THE BURGH OF MAYBOLE."—*Present seal of the burgh.—Communicated by E. Colquhoun, Esq.*

1243. MONTROSE. A royal burgh in Forfarshire.

The Crucifixion.

"SIGILLUM COMUNE [DE] MUNRO."

1244. COUNTER SEAL OF THE LAST.

A rose. Both these are very rudely executed.—*Appended to the same Instrument as No. 1197.*

1245. NAIRN. The capital of the county, and a royal burgh at an early period.

A full-length figure of a saint, crowned with the nimbus, a cross staff in his right hand, and an open book in his left.

"SIGILLUM COMMUNE BURGI DE NAIRNE."—*From the original copper matrix in the office of the Town-Clerk of Nairn.*

On the handle of the seal is the following inscription, "EX DONO JOANNIS ROSS DE NEUCK REPRESENTATIVE IN PARLIAMENT AND BURROWS FOR NAIRN, 1703,"—the words, "Convention of the Royal" require to be inserted before "Burrows" to make the sense clearly apparent.

1246. NEWTON-ON-AYR, AYRSHIRE. A royal burgh, rather singularly constituted; with the burgh of Prestwick, it received many privileges from King Robert I., which were all confirmed by King James VI. A.D. 1595. *Plate XIV. fig. 3.*

A fess chequé between three monograms—; a slight foliage surrounds the shield.

"SIGILLUM COMUNE NOVE VILLE DE ARE."—*This matrix is a good example of fifteenth-century work, and is still used by the burgh.*

1247. PERTH. The capital of the county, and probably created a royal burgh by William the Lion.

A figure of St. John the Baptist, "in raiment of camel's hair," holding the Agnus Dei; on each side are two monks kneeling at prayer; all the figures are within niches or the porch of a church.

" S̄ COMUNITATIS VILLE SANCTI JOHANNIS BAPTISTE DE PERTH."

1248. COUNTER SEAL.

The decollation of St. John; Herodias waiting with a charger. The legend is the same as on the last.— *Appended to the Homage Deed of the Burgh, 1296. —H.M. Record Office.*

A very fine example of the art, and fortunately in excellent preservation. This seal is also appended to "Obligation of

the Burghs of Aberdeen, Edinburgh, Perth, and Dundee," A.D. 1423, but the

obverse is evidently from a different matrix. It is a very close copy, but deficient in the spirit of the earlier one, which has probably been lost or broken at some period subsequent to 1295-6.

1249. PITTENWEEM, Fife.   Erected into a royal burgh by King James V., A.D. 1537.
A figure of a bishop standing in a boat rowed by two naked boys; at the stern a flag bearing a lion rampant.
"SIGILLUM COMMUNE BURGI DE PITTENTEM."—*Matrix in the office of the Town-Clerk of Pittenweem.—Communicated by E. Colquhoun, Esq.*

1250. PRESTWICK, Ayrshire.   A very ancient burgh of barony, possessing some singular privileges and customs.   *Plate XIV. fig. 7.*
A full-length figure of a bishop (St. Ninian?) within a niche, mitred, and crowned with the nimbus, his right hand bestowing the benediction, his left holding the crosier; at the dexter side of the niche is a castle with lofty central tower, and at the sinister side a branch of foliage.
"SIGILLUM COMUNE BURGI DE PRESTVIC."   This, like the matrix of Newton, is fifteenth-century work, and a good seal, still used by the burgh, which may share, with Newton and Forres, the credit of having taken greater care of their ancient seals than most of the burghs of Scotland.—*Communicated by E. Colquhoun, Esq.*

1251. QUEENSFERRY (South), Linlithgowshire.   A royal burgh.
A full-length figure of St. Margaret with a sceptre in her right hand; in the lower part of the seal a small ship.
"S' COMMUNE BURGI PASSAGII [REGINE]."—*Appended to an Instrument dated 21st January 1529, belonging to the Incorporation of Tailors of Queensferry.*

1252. QUEENSFERRY (South).
A figure of St. Margaret standing in a boat.
"INSIGNIA BURGI PASAGI REGINE."

1253. COUNTER SEAL of the last.
On a cross fleury, between four martlets, one of the same, not on a shield.—*Matrix in the office of the Town-Clerk.—Communicated by E. Colquhoun, Esq.*

1254. ROTHESAY. The capital of the county of Bute, and a royal burgh, created by Robert III., A.D. 1401.

Per pale. Dexter, a castle between a crescent and a mullet in chief, and a galley in base; sinister a fess chequé. (This is evidently a modern composition of the fine old seal of the burgh, Nos. 1179-80 of "Descriptive Catalogue of Ancient Scottish Seals.")

" LIBERTAS DATUR VILLE DE ROTHISEA."—Communicated by E. Colquhoun, Esq.

1255. ROXBURGH. A royal burgh in Roxburghshire, created by King David I. before A.D. 1150.

A fine seal. The arms of Scotland suspended on a tree, and on each side an eagle on the branches.

" [S] COMUNE BURG[I] DE ROKESBURC."

1256. COUNTER SEAL.

A tower, from the upper battlements of which a shield, bearing the arms of Scotland, is suspended. From the side turrets a warder is sounding an alarm. In the gateway is a pilgrim, and two crowned heads appear at the windows.

" . . . CHASTEL R MOIT . . . RCI EMND ESTAPE IEOE VE ESTEM."— Appended to the Homage Deed of the Burgh, 1296.—H.M. Record Office.

1257. STIRLING. The capital of the county, and among the earliest royal burghs.

A lamb conchant on the top of a rock.

" OPPIDUM STERLINI."—Communicated by E. Colquhoun, Esq.

1258. THURSO. CAITHNESS. Erected a burgh of barony by King Charles I., A.D. 1633.

A figure of St. Peter, with the keys in his right hand and a patriarchal cross in his left.

" SIGILLUM BURGI DE THURSO, IN CAITHNESS."—An exceedingly rude seal.—Communicated by E. Colquhoun, Esq.

1259. TAIN. A royal burgh, and the county town of Ross-shire.

A figure of St. Duthac, a staff in his right hand and an open book in his left. On a ribbon is inscribed " TANTE SIGILLUM COMMUNE." and in an inner circle, " S' BEATUS EST DUTHACUS."—Present seal of the burgh.—Communicated by E. Colquhoun, Esq.

1260. WHITHERN, Wigtonshire.   Candida Casa.   A royal burgh at an early period.
        Plate XIII. *fig.* 3.

> A figure of a bishop (St. Leonard), mitred and crowned with the nimbus, sitting,
> his right hand on his breast, his left resting on a book upon his knee.   At
> each side are three links of a chain.   Background foliated.

> " S' COMUNI CIVITATIS BURGI CANDIDE CASE.—*Matrix in the office of the Town-
> clerk of Wigton.—Communicated by F. Colquhoun, Esq.*

1261. WHITHERN and WIGTON.

> A figure of St. Leonard, his right hand raised, a chain and fetterlock suspended
> from the wrist.   In his left hand the crosier.

> " S' . . . . . QUHITHYRNE ET VIGTOUNE."

> The original matrix of this seal is in the office of the Town-clerk.   It is of copper,
> and is evidently the matrix of only one side of the seal, having four perfor-
> ated projections for receiving the pins of the matrix of the opposite side,
> which is wanting.   It is a very extraordinary seal, combining the names of
> two burghs, which, so far as any evidence can be found, were never united
> either as a municipal or ecclesiastical body.   The word before Quhithyrne
> has obviously been blundered by the engraver.   It is impossible to say what it
> is meant for.

APPENDIX.

I.—ADDITIONAL SEALS.

1262. ALEXANDER, SIR WILLIAM, AFTERWARDS EARL OF STIRLING.

Per pale; a crescent. Crest, two hands joined, holding a branch of foliage. Perhaps this latter is only an ornament, forming no part of the crest.— *From a Letter to the Viscount of Stormonth, dated 23d November* (1625).— *David Laing, Esq.*

1263. ARRAT, GEORGE, "OF THAT ILK."

A cross-crosslet, fitchée, between two mullets.

"s' GEORGE ARRAT."—*Appended to an Inquisition, dated* A.D. 1523.—*Crawford and Balcarras Charters.*

1264. ARTHUR, JAMES.

A chevron between three mascles in chief, and a cross-crosslet, fitchée, in base. Foliage surrounds the shield.

"SIGILLUM JACOBI ARTHUR."—*Appended to Charter of Alienation by James Arthur, son of William Arthur, to John Learmonth, 4th April* 1457.—*St. Mary's College Charters, St. Andrews.*

1265. ARTHUR, ROBERT.

A peacock's head couped, in chief, and a mascle in base on the dexter; and on the sinister two mascles, palewise.

"s' ROBERTI ARTHUR."—*Appended to Sasine of a tenement, dated* A.D. 1551.—*St. Andrews Charters.*

2 F

1266. BALBERNIE, ALEXANDER, of Innerrichtie.

A lion rampant.

" s' alexandri d balbirne."—*Appended to Reversion or Discharge by Alexander Balbernie and his Son, to Mr. John Lindsay, parson of Menmuir, of the sunny half of the lands of Balmakin, 12th May 1589.—Crawford and Balcarras Charters.*

1267. BALDEVI, DAVID.

Three geese passant.   Foliage at the top and sides of the shield.

" s' davidis baldevi."—*Appended to Procuratory of Resignation by D. Baldevi, burgess of Montrose, in favour of Sir David Lindsay of Edzell, Knight, of the Temple lands of Newdock, etc., A.D. 1594.—Crawford and Balcarras Charters.*

1268. BONAR, JOHN, of Rossy.

A chevron between two crescents in chief and a mullet of six points in base.

" s' johannis bonar."—*Appended to a Reversion of the lands of Cocklaw by John Bonar of Rossy, 12th January 1509.—St. Salvator's College Charters.*

1269. BONAR, WILLIAM, of Rossy.

A saltire couped.   In base a crescent.

" s' wilmi bonaer."—*Appended to Precept of Sasine of the lands of Carnbody, co. Perth, in favour of his eldest son and heir, John Bonar, 20th October 1549.—H.M. General Register House.*

1270. BONAR, WILLIAM.   The same person as the preceding.

Quarterly: first and fourth, a saltire couped, in base a crescent; second and third, on a fess, a mullet.   Foliage at the top and sides of the shield.

" s' villelmi bonar de rosy."—*Appended to Resignation of the lands and barony of Carnbody to himself and Joana Johnston, his spouse, 7th August 1579.—H.M. General Register House.*

1271. CARMICHAEL, JOHN.

A fess wreathed.

" s' johan de carmichael."—*Appended to an Indenture between John Carmichael and John of Kinlochy, 11th February 1434.—St. Salvator's College Charters.*

1272. CARMICHAEL, WILLIAM, of Meadowflat.

A fess wreathed.

" s' wilmi de carmichael."—*Appended to Charter by William Carmichael to Gilbert Kennedy of Kirkmichael, of a tenement in St. Andrews, A.D. 1462.—St. Salvator's College Charters.*

1273. CASS, THOMAS.

A signet. On a saltire, between four cross-crosslets, fitchée, five crescents. Crest on helmet, and mantlings, a cross-crosslet issuing from a crescent.—*From a Letter to the Earl of Ancrum, dated 6th May 1637.—Marquis of Lothian.*

1274. CHARTERIS, HENRY, Bailie of Edinburgh. The eminent bookseller and printer in Edinburgh (1568-1599).

On a fess, an annulet, within the royal tressure.

" s' HENDRICI CHARTRIS."—*Appended to an Instrument of Sasine dated 1590.— David Laing, Esq.*

1275. CLAPEN (or CLEPHANE), ANDREW.

Per pale; dexter, a chevron between three mascles; sinister, a helmet in profile, and on it, for crest, a boar's head. Foliage at the top and sides of the shield.

" s' ANDRII CLAPEN."—*Appended to Charter by Andrew Clapen, portioner of Pitcorthries, of an annual-rent of forty merks furth of the lands of Wester Pitcorthries, to Mr. John Lindsay, parson of Menmuir, 26th May 1558.— Crawford and Balcarras Charters.*

1276. CRAMOND, JAMES, of that ilk.

On a bend, sinister (doubtless a mistake of the engraver), three pelicans.

" s' JACOBI CRAMONT."—*Appended to Reversion by James Cramond to David Earl of Crawford of the lands of Dulquharth, in the barony of Glenesk, 25th August 1522.—Crawford and Balcarras Charters.*

1277. CROUSTON, GILBERT.

Three mullets. Foliage at the top and sides of the shield.

" s' GILBERT CROUSTOUN."—*Appended to Charter by James Broun, son and heir of the late Robert Broun of Keir, to Ronald Masterton of Bad, of the lands of Keir, in the lordship of Culross and sheriffdom of Perth, 22d January 1573. —Principal Campbell, Aberdeen.*

1278. CUNINGHAM, SIR WILLIAM, Knight, afterwards fourth Earl of Glencairn.

A pall. Undoubtedly meant for the shakefork.

" GUILLERIMUS CONYNGHAM MILES."—*Appended to Assignation by Sir William Cuningham, Knight, Master of Glencairn, to his son and heir-apparent, Alexander Cuningham (afterwards fifth Earl of Glencairn), to the reversion of eight hundred merks gold and silver, upon the redeeming of the lands of Dreyhorn and others, within the barony of Reidhall and sheriffdom of Edinburgh, 23d October 1531.—David Laing, Esq.*

1279. DEMPSTER, JOHN, of CARALDSTON.

    A lion rampant, debruised with a ribbon.

    " s' JOHANNIS DEMPSTAR DE CARALDSTON."—*Appended to Precept of Sasine by John Dempster in favour of Sir David Lindsay of Ezdell, Knight, of the lands of Brocklaw, 13th June 1509.—Crawford and Balcarras Charters.*

1280. DALRYMPLE, JAMES, FIRST VISCOUNT STAIR.

    A signet. Quarterly: first and fourth, on a saltire, nine lozenges, for Dalrymple; second, a chevron chequé between three water-bougets, for Ross of Balniel; fourth, a lion rampant, for Dundas of Newliston. A label of three points. Above the shield a coronet of nine points. Supporters, two storks, each with a serpent (eel ?) in its beak. Below the shield, "QUIESCAM."—*From a Letter to the Earl of Lothian, dated 16th January 1692.—Marquis of Lothian.*

1281. DRUMMOND, JANE, COUNTESS OF ROXBURGH.

    A very pretty octagon-shaped signet. Three bars wavy. The shield between two branches of laurel, and an earl's coronet above it.—*From a Letter in French, signed " Jane Roxbroughe," to the Prince of Orange, dated at The Hague, 7th July 1642.—David Laing, Esq.*

1282. ERSKINE, JOHN, EARL OF MARR. REGENT OF SCOTLAND, A.D. 1571.

    Quarterly: first and fourth, a bend between six cross-crosslets, fitchée, for Marr; second and third, a pale, for Erskine. Above the shield a coronet of nine points.— *From a Letter to Lord Gray, dated at Leith, 11th September 1571. —David Laing, Esq.*

1283. ERSKINE, ALEXANDER, THIRD EARL OF KELLIE.

    A signet. Quarterly: first and fourth, an imperial crown within the royal tressure, as a coat of augmentation, for the title of Kellie; second and third, a pale, for Erskine. Above the shield, an earl's coronet. Supporters, two griffins.— *From a Letter to the Earl of Linlithgow, dated 22d June 1608.—David Laing, Esq.*

1284. GRAHAM, JOHN, EARL OF MONTROSE. LORD TREASURER OF SCOTLAND, A.D. 1584; LORD CHANCELLOR 1598.

    A small signet. Quarterly: first and fourth, on a fess, three escallop shells, for Graham (a chief charged with three escallop shells is the correct blazon); second and third, three roses, for title of Montrose. The shield is placed between the initials I · M, with an earl's coronet above it.—*From a Letter to the King (James VI.), dated 20th January 1603.—Sir James Balfour's Collection, Advocates' Library.*

1285. GRAHAM, WILLIAM, Earl of Airth and Menteith.

A very well executed signet. Quarterly: first and fourth, on a chief, three escallop shells, for Graham; second and third, a fess chequé, for Menteith; and in chief a chevron, for Strathern. Above the shield a coronet, and around it the the initials w · m (William Earl of Menteith).—*From a Letter, signed "Airthe," to the Laird of Gairtmoir, dated 10th March 1663.—David Laing, Esq.*

1286. GUTHRIE, JAMES. Minister of Stirling, the author of "*The Causes of God's Wrath;*" executed at Edinburgh A.D. 1661.

A small signet. Shield quarterly: first and fourth, two garbs, for Cumin (?); second and third, a lion rampant, for Guthrie. Above the shield the initials I · G (Mr. James Guthrie).—*From a Letter to Lord Warriston, dated 23d October 1659.—David Laing, Esq.*

1287. HAMILTON, SIR THOMAS. *Vide No. 472.*

A very neat signet. On a chevron, between three cinquefoils, a buckle, tongue erect. Above the shield the initials T · H (Sir Thomas Hamilton).—*From a Letter to the King (James VI.), dated 4th June 1609.—Sir James Balfour's Collection, Advocates' Library.*

1288. HEPBURN, JAMES (of the family of Blackcastle). Dean and Vicar-General *sede vacante* of the Cathedral Church of Dunkeld.

On a chevron, a rose between two lions rampant, respecting; in base a heart-shaped buckle. Foliage surrounds the shield.

"s' jacobi hepburn vicarii generalis dunkelden."—*Appended to Precept of Sasine, 18th February 1549.—David Laing, Esq.*

1289. KER, ROBERT, first Earl of Roxburgh.

A signet. On a chevron, between three unicorns' heads erased, as many mullets. An earl's coronet above the shield, and around it the initials R · R (Robert Earl of Roxburgh).—*From a Letter to the Viscount Annan, dated 26th July 1623.—Sir James Balfour's Collection, Advocates' Library.*

1290. LINDSAY, JOHN, sixth Earl of Crawford.

Quarterly: Lindsay and Abernethy.

"s' johannis comitis crawfurd."—*Appended to Precept of Sasine of the lands of Carnbody, in sheriffdom of Perth, in favour of John Bonar, eldest son and heir to James Bonar of Rossy, 2d December 1512.—H.M. General Register House.*

1291. MONTGOMERY, HUGH, sixth Earl of Eglinton.

A signet. A very pretty, but quite a fanciful arrangement of the
Eglinton and Montgomery quarters. Three annulets gemmed,
a fleur-de-lis in each. Above the shield an earl's coronet.—
From a Letter to John Murray, of H.M. Bed-Chamber, A.D
1620.—*Sir James Balfour's Collection, Advocates' Library.*

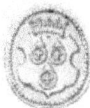

1292. ROSS, THOMAS, Commendator of Ferne (Ross-shire) and Provost of Tain.

Three lions rampant. Foliage at the top and sides of the shield.

" S' MAGISTER THOME ROS PREPOSITI DE TAIN."—*Appended to the same Instrument
as No. 1143.*

II.—UNKNOWN SEALS.

1293. A knight on horseback, armed in edged ring-mail, in his right hand a broad-bladed
sword, and on his left arm an uncharged shield, with the umbo or boss in the
centre.

" SIGILL. BENEDICAMUS DEI ANNU FLIL." (?) Some of the letters are reversed and
some inverted.

This is a very extraordinary seal or matrix, being a flat disk of brass, rather thicker
at one side than the other, and on the back are some foliated ornaments
engraved, but without any design. They seem mere trials of the tool. There
is no appearance of any kind of handle having been on it. The style of the
whole is that of the twelfth century, though the rowel spur might raise a
suspicion of its being something later.—*Found among some old metal at Rae-
wick, in Shetland, and purchased by the Society of Antiquaries of Scotland.*

1294. A male head, with a turban. Profile to dexter, and a
branch of hyssop (?) or laurel in front. Inscription
in Hebrew characters.—*Brass matrix, found on the
east side of Arthur Seat, near Duddingston.—
Museum of Society of Antiquaries of Scotland.*

Some explanation of this, and notes of other seals with
Hebrew legends, will be found in the " *Proceedings
of the Society of Antiquaries of Scotland*," vol. i. pp. 39-150.

1295. LODENCOURT, JOHN.

> Couché, an eagle displayed, surmounted by a bend. Crest, on a helmet, a dragon
> vomiting flames.
>
> "s' JEHAN DE LODENCOURT."—*Brass matrix, found in the Molendinar Burn, Glas*
> *gow, now in the Museum of the Society of Antiquaries of Scotland.*

1296. AYSELL, EUSTACE.

> The Agnus Dei.
>
> "s' EUSTACHII DE AYSELL."—*Brass matrix, found in Aberdeenshire, now in the*
> *Museum of the Society of Antiquaries of Scotland.*

1297. A three-sided figure, with a circle, or ring, at each angle.

> "s' LVEN CIBIN ROLLSSE." It has been read, "s' IVAN CLOM
> ROLLNE." But neither is satisfactory.
>
> This is a very singular device or emblem—it may be of the
> Holy Trinity As an example of fourteenth-century work
> it is very interesting; and a better explanation both of the device and legend
> is very desirable.—*The matrix, of brass, was found in repairing an old house*
> *in Lanark, and is now in the possession of Adam Sim, Esq., of Coulter Mains,*
> *Lanarkshire.*

1298. A most singular design, which may perhaps be meant to represent a burning rock,
to which, at first sight, it bears some resemblance ; but on closer examination
it appears more like water flowing into and overflowing a square vessel.

> "s' DNI VIANESII D LA VELOLONGO."—*Brass matrix, for many years in the pos*
> *session of a family in Edinburgh, now in that of James Macdonald, Esq.,*
> *Edinburgh.*

1299. ST. ANDREWS, THEOLOGICAL COLLEGE of.

> A rudely-executed design. An allegorical figure of religion. A winged female,
> leaning against a Tau cross, holding in her right hand an open book, extended
> towards heaven, standing in a crescent upon a skeleton (emblem of death).
> From the transverse limb of the cross is suspended a bit and bridle. Rays
> intended perhaps for an aureole, surround the design.
>
> "s' COLLEGII ST. THEOLOGIE DICATI F' ANDRE. RELLIGIO SUMMI SANCTA PATRIS
> SOBOLES."—*From the original brass matrix in St. Mary's College, St. Andrews.*
>
> It was for some time supposed that this was the seal of St. Mary's College, but it is
> more probably that of some continental college dedicated to St. Andrew ;

indeed, this is placed beyond a doubt by the fact that the same device is found as a printer's device, but without the inscription, on three examples, printed respectively at Lyons A.D. 1565, Geneva A.D. 1618, and Sedan A.D. 1633. The two former are Calvin's "*Institutes*," and the latter a Bible, where it appears as a vignette on the title. In the Lyons edition of Calvin the design is accompanied with explanatory rhymes, entitled, "*Pourtrait de la Vraye Religion.*"

It may deserve mention that a copy of the Bible referred to was once in the possession of the celebrated Marquis of Montrose, and some highly interesting notes regarding its history, and a lithograph copy of the title-page, will be found in Mr. Napier's "*Life and Times of John Graham of Claverhouse,*" 1859, vol. i., Introduction, p. xxx. We are indebted to Mr. David Smith, St. Andrews, for much of the information regarding this curious seal. But all claim for its being a Scottish seal must be given up.

III.—DETACHED SEALS.

THE following are all from detached seals preserved in H.M. Record Office. They were undoubtedly appended to various documents relating to the affairs of Scotland, and the prevailing style of art in which they are executed connect them with the thirteenth and fourteenth century. Many, indeed, would indicate even an earlier period. Where the device and a motto only appear, it is impossible to identify them or assign them to any particular owner. It has therefore been thought best to form these into a distinct class, to follow those having a name on them, the ownership of which is thereby ascertained. The majority of the seals in both classes are merely devices, not upon a shield, and therefore not perhaps strictly heraldic. Where the shield does appear it is specially mentioned.

1300. GILBOHIN, PATRICK.
A rose.   " s' PATRICII D GILBOHIN."

1301. GILMOYAN, JOHN.
A star, or wheel ornament.   " JOHIS GILMOYAN."

1302. GRAYDN, WALTER DE.
A crown of thorns. Very prettily executed.   " S WALT. D. GRAYDN."

1303. HAMENONE, JOHIS DE.
>A flower or wheel ornament. "s' JOHIS DE HAMENONE."

1304. HANMEL, GILBERT.
>A lion rampant to sinister. "s' GILBERTI HANMEL."

1305. HATTER, WILLIAM
>Oval shape. A man's head in profile to left; above it a cross.
>"s' VILLAMI HATTER."

1306. KILMEROC, REGINALD DE.
>A flower or wheel ornament. "s' REGINALD D. KILMEROC."

1307. KIMOREN, WILLIAM DE.
>An archer shooting a stag. "SIGILL. WILLI DE KIMOREN."

1308. KYLTON, SIMON DE.
>The martyrdom of St. Thomas à Becket. "s' SIMONIS DE KYLTON."

1309. LEMESSAG, MICHAEL.
>A cross between four mullets of six points. "s' MICHAEL LEMOSAG"

1310. LIPE, JOHN.
>A dog coiled. "s' JOHANNIS LIPE"

1311. MALHERB, GILBERT.
>A bunch of herbs. "s' GILL. MALERBE."

1312. OTYR, JOHN.
>Two chevrons. Above the shield the head of a spear
>"s' JOHANNIS OTYR."

1313. PARD, JOHN.
>A stag running. A bow and arrow at the dexter side.
>"s' JOHANNE PARD (?)."

1314. PAYNERE, RALPH.
>A singular device of a fox or wolf on its hind-legs, holding a shepherd's crook in
>its fore-paws; in front is a rabbit, dog, and bird.
>"s' RAULPHI PAYNERE (?)."

1315. PEACOCK, HUGH.

Oval shape. A peacock walking. " S' HUGONIS PECOK."

1316. PINKERTON, NICOLAS.

A dog running, and above it a rose.
" S' NICHOLAI DE PINGKERTON.".

1317. PLEMING, JOHN.

A bend; in sinister chief a mullet of six points. At the sides of the shield two
lizards, and at the top foliage.
" S' JOHANNIS PLEMING."—*Melrose Charters.*

1318. PRESTON, WILLIAM.

A cock and a flower. " S' WILLI D. PRESSTON."

1319. PRINGLE, HELIAS.

Oval shape. A hunting-horn. " S' HELIAS DE HOPPRIEGIL."

1320. RAIT, JOHN.

A hunting-horn. " JOH. DE RAIT."

1321. RAMSAY, ADAM.

Octagon shape. A hawk resting on a hand; in the background three roses.
" S' ADE RAMSAY."

1322. RAMSAY, JOHN.

A well-executed device of a falcon in the midst of foliage.
" S' JOHANNIS DE RAMESAYE."

1323. RAMSAY, WILLIAM.

An eagle with wings expanded.
" S' WILLI DE RAMESEYE."

1324. ROLLOK, JAMES.

A chevron between three lions' heads erased, and a mullet for difference.
" S' JACOBI ROLLOK."

1325. RENART, RICHARD.
   Oval shape. A curious device of a fox (Reynard) on its hind-legs, holding in its
      fore-paws a pair of scales.
   "S' RICHART RENART."

1326. RIMUR, ANDREW.
   A pelican feeding her young.   " S' ANDREE FIL RANULFI RIMUR."

1327. RISIN, BARTHOLOMY.
   A star or wheel ornament.   " S' BARTHOL. DE RISIN."

1328. SERVITORIS, WILLIAM.
   Oval shape. The Virgin and Child sitting.   " WILLI SERVITORIS."

1329. TOMSON, WILLIAM.
   Oval-shaped seal. A hand above a chalice.   " S' WILLI TOMSON CAPELL."

1330. VAPRER, WILLIAM.
   A wheel ornament.   " S' WILL. VAPRER."

1331. VITALI, JOHN.
   Oval shape. A squirrel on a tree.   " S' JOANIS VITALI."

1332. WALE, THOMAS.
   A star.   " S' TOME WALE."

1333. WALRAN, PHILIP.
   A flower or wheel ornament.   " S' PHILIPPI DE WALRAN."

1334. WALTER, WALTER, SON OF.
   Device of a fox or dog; a crescent, star, and flower.
   " S' WALTERI FILLI WALTERI."

1335. WATT, JOHN.
   An eagle.   " S' JOHANNIS LE WAAT."

1336. WHITELAW, JOHN.
   A chevron between three boars' heads.   " S' JOHANNIS DE QUHITLAW.

1337. WIPOND (?) ROBERT.
    A fleur-de-lis.   " s' ROBERT WYPPUNT "

1338. WITON, ADAM.
    A wheel ornament. " s' ADE DE WITON."

1339. WISSAIT, GILBERT.
    A dog's head. " s' GIBBERT WISSAIT."

1340. WYNHOU (?), JOHN.
    A lion combating a dragon.   " s' JHONNAI DE WYNHOU" (?).

1341. " ESTO FIDELIS."   A crescent and a star.—*Appended to the same Homage as No. 18, A.D. 1296.—H.M. Record Office.*
    This is doubtless the seal of some Crusader with whom the crescent and star was a favourite device.

1342. " ICE PAS CO." (?)   A device of a lion coiled in centre of two squares interlaced.—*Appended to the same Homage as No. 18, A.D. 1296.—H.M. Record Office.*

1343. " CREDE MICHI."   A rose.

1344. " MINA SECRETA TEGO."   An antique gem. A head in profile to sinister—*Appended to Cirograph, dated 7th July 1292.*

1345. " LE SEEL FERGUS "   A front male head between two fleurs-de-lis.

1346. " s' ROBERTI CAPLI "   A hand above a chalice.

1347. A squirrel between two fleurs-de-lis, and above is the word " PKIUS," on a shield-shaped seal.

1348. " s' NICHOLAI CAPELLANI."   Oval. A priest consecrating the chalice.

1349. " ICRAKENOTTIS" (I CRACK NUTS).   A squirrel.

1350. " CAPUD STEVI DEI."   A male head, profile to sinister.

1351. " SIGILLUM SERAI MARIE." An antique gem. A figure surrounded by foliage.

1352. " AVE MARIA GRACIA." A boar passing to the sinister, in front of a tree.

1353. " S' MULLUM TALE (?)." An ass on its hind legs, carrying a head or mask in its fore-legs.

1354. " SIGILL. AMORIS." A griffin walking.

1355. " CRISMO . . . CLASS ROT CAYCOU MASSE." A man riding on a donkey, hawking.

1356. " TIMETE DEUM" ? . A pelican feeding its young.

1357. " SOHO WHOU . . . S. A dog passing in front of a tree.

1358. " CREDE MICHI " A bird standing.

1359. " SIGILLUM SECRETI." A lion rampant.

1360. " TENET MONCHAIRE TEMANOUS " (?). Two demi figures of angels holding up a wreath.

Nº 6.

3

4

5

6

5

8

4

5

7

8

3

4

5

6

8

www.ingramcontent.com/pod-product-compliance
Lightning Source LLC
Chambersburg PA
CBHW021511210326
41599CB00012B/1220